BY FRANK J. ROMANO

MACHINE WRITING AND TYPESETTING

The story of Sholes and Mergenthaler
and the invention of the
typewriter and the linotype

"I want to bridge the gap between the typewriter
and the printed page."

James O. Clephane, 1876

GAMA

For Joanne
. . .who had the good sense
to marry the mail boy.

© 1986 Frank J. Romano
All rights reserved

Printed in the United States of America

Library of Congress Catalog Number 86-81272
ISBN: 0-938853-00-7

GAMA, PO Box 170, Salem, NH 03079, (603) 898-2822

TABLE OF CONTENTS

CHAPTER 1
The magic clock, the spader and the writing machine ...... *Page 1*

CHAPTER 2
The day the clock's hands moved .... *Page 17*

CHAPTER 3
The transfer typewriter .... *Page 23*

CHAPTER 4
The impression machine .... *Page 29*

CHAPTER 5
The band machines .... *Page 39*

CHAPTER 6
The independent matrix machine .... *Page 49*

CHAPTER 7
The publishers syndicate .... *Page 57*

CHAPTER 8
The Mergenthaler Line-of-Type Machine .... *Page 65*

CHAPTER 9
The eighth wonder of the world .... *Page 81*

CHAPTER 10
The son who did not become a teacher .... *Page 97*

Chronology .... *Page 107*
Bibliography .... *Page 117*
Index .... *Page 121*

FOREWORD

This is the story of the men and the machines involved in the development of mechanical approaches to writing and typesetting. Both the typewriter and the Linotype have played important roles in written and printed communication, and the whole story of their genesis, with all of its intricate relationships, has never been told.

The impetus for this book comes from the celebration of the centennial of the introduction of the Linotype, at a time when laser printing and electronic workstations are changing—again!—the way we work with words.

The prophetic words of James Clephane about linking the origination capability of the typewriter with the reproduction capability of the printing process have important meaning today as we struggle to harness technology—electronic instead of mechanical—to the publishing task.

It is altogether fitting and proper that we try to understand the evolution of the processes and tools of the printed page, and to know the people who, in many respects, changed history. As you read about the past, you may find interesting parallels to the present—and probably the future.

This book could not have been written without the contribution of many people, who were willing to share their time and experiences, as well as searching long-forgotten cartons of books and records. I am especially grateful to John Schappler for organizing and checking the manuscript and Marsh Brown for tracking Sholes through Wisconsin.

John Kennedy was about to become president when I went to work for Mergenthaler Linotype Company at 39 Ryerson Street,

Brooklyn, New York. My first job was delivering the mail, a task that brought me into contact with all parts of the firm and provided me with an opportunity to see the breath of the company and its products. I would often stand in awe when Herman Mergenthaler would visit the company; eventually I would get the nerve to ask him questions.

There was a wealth of historical information there then. The "museum" of historic typesetting machines was there, eventually donated to different collections; the Blower model in the 7th floor lobby later went to the Smithsonian. The bust of Ottmar, on which I used to hang flowers and hats, may still be there.

This may be the last chance to tell the story of mechanical technology. There are still enough of us that remember it to care. The next wave of historical writings will certainly consider photographic typesetting and its evolution into electronic approaches.

The second hundred years are never as important as the first.

Frank Romano
Salem, New Hampshire

CHAPTER 1

THE MAGIC CLOCK, THE SPADER AND THE WRITING MACHINE

The fifty-second person to invent the typewriter, and the only one to call it that, was Christopher Latham Sholes.

It was February 14, 1819 when the Sholes family announced from their log farmhouse near the village of Mooresburg in Montour county, Pennsylvania the arrival of a son named Christopher Latham. He grew up in Pennsylvania and worked with his father, a cabinetmaker.

Mary Jane McKinney, a boyhood sweetheart, went to the same rural school and they took turns playing on each other's farms and promising that when they grew up they would marry each other. While still young his parents moved to Danville, Pennsylvania where his mother died in 1826, when the Christopher was seven.

Christopher was apprenticed to the printing trade as a "shop devil" on the *Democratic Danville Intelligencer,* founded in 1828. After four years' apprenticeship, his father moved to Wisconsin, settling in Green Bay in 1837. His brother, Charles C., a printer like himself, owned and edited the *Green Bay Democrat.*

Land offices had been established at Green Bay as well as Mineral Point in 1834 as a result of President Andrew Jackson's public land sale proclamation. With the opportunity of owning portions of the state, the vanguard of early settlers built homes on the sites of the present cities of Kenosha, Milwaukee, Sheboygan, and Racine, starting in 1835.

At twenty, in 1839, he left home to follow his brother Charles to Madison where the latter had acquired an interest in the *Wisconsin Enquirer.* Christopher was given charge of the paper while continuing to supervise public printing.

After service for a year as its editor and as journal clerk of the legislature, he moved to Southport (later Kenosha), Wisconsin to establish, with friend Michael Frank, the *Southport Telegraph*, afterwards the *Kenosha Telegraph*, a name inspired by Samuel F. B. Morse's invention.

Morse's demonstration of his magnetic telegraph inspired Sholes and other newspaper people about its importance in the rapid transmission of news. He was twenty one in 1840, public spirited and honest, and soon became a trusted leader in Kenosha. He was appointed Southport's postmaster in 1843 by President Polk, and after three or four years' service became postmaster of Milwaukee also.

As he promised, he married Mary Jane McKinney in Green Bay on February 4, 1841 and was the father of ten children, five boys and five girls.

It was usual for newspapers to conduct a job printing department for additional income and, one year, the compositors on Sholes' newspaper went on strike. This so angered Sholes that he gave serious thought to typesetting by machine. Being a tinkerer at heart, he built models in which types could be impressed in wax, but the wax bulged and castings of molten metal were not useable. He cast his models aside and made peace with the staff.

His career moved back and forth between publishing and public service. He was state senator from Racine county in 1848 (the year Wisconsin was admitted to the Union) and 1849, and from Kenosha county in 1856 and 1857, He had previously represented Kenosha in the assembly in 1852 and 1853. Sholes, at the age of forty eight in 1867, was a man with a distinguished career as newspaperman, printer and politician. He was a pleasant, congenial, friendly gentleman given to smiling, though not to boisterous laughter.

As editor he always printed adverse criticisms about himself by his political adversaries, while at the same time omitting complimentary mentions of his work or himself. He seemed to go through life doing what he did and drawing little attention to himself. Even his clothes were plain and he was inclined to be a little careless about his dress. His appearance was said to be that of a poet's, which may have been a kind way of saying something unkind. He was tall, almost lanky, with wild flowing hair and a scrappy beard. He loved puns and his idea of the world's best joke was a poetic pun. He came home one night, for example, and found a kerosene lamp had dripped onto the rug in the hall. Shole's comment, from Byron, was: "The isles of Greece, the isles of grease! Where burning Sappho loved and sung . . ."

In his whole lifetime he never sought nor obtained rewards or honors and few people associate his name with one of the significant contributions to print communication.

Sholes and his friends in Milwaukee spent their spare time inventing things, tinkering with carpentry and shop work. Their unofficial headquarters was a machine shop run by C. F. Kleinsteuber in a small wooden building on the north edge of Milwaukee. Sholes had devised a method for addressing newspapers by printing names of subscribers on the margin, as well as tinkering with other gadgets.

It had been his job, as a lad, to number by hand the pages of account books with a metal stamp. The use of a metal "finger" for this purpose was a much better idea and it occurred to him that he could devise a machine to perform this work much more neatly and quickly.

He discussed this project with his friend Samuel W. Soulé and they began work in a small dingy room on the upper floor of the old mill. In another workshop Carlos Glidden, an attorney, became interested in the experiment and offered his talents. He was developing a mechanical "spader" that would take the place of the plow. The trio discussed their plans and debated the weak points as their experiments progressed. Dr. Henry W. Roby, a court reporter working on a trick clock to be used in a magic act, was also there that summer and joined them on occasion.

Referring to the page numbering machine, Glidden asked: "Why can't such a machine be made that will write letters and words and not figures only? Sholes, why can't you build a machine to print letters and words as perfectly as these figures are struck off here?"

During that eventful summer of 1867, the group obtained a description of a writing machine from *Scientific American* magazine (July 6, 1867 issue) which quoted an article from a London technical journal about a machine invented by American John Pratt, of Centre, Alabama. It was certainly a roundabout way to get an idea. The article described experiments with the "pterotype," a name translated as "winged printer".

"A machine," said editor Alfred Ely Beach, in the issue, "by which it is assumed that a man may print his thoughts twice as fast as he can write them and with the advantage of the legibility, compactness, and neatness of print, has lately been exhibited before the London Society of Arts, by the inventor, Mr. Pratt, of Alabama. The subject of typewriting is one of the interesting aspects of the near future. Its manifest feasibility and advantage indicate that the laborious and unsatisfactory performance of the pen must, sooner or later, become obsolete for general purposes. "Printed copy" will become the rule, not the exception, for compositors, even on original papers like the Scientific American. Legal copying, and the writing and delivering of sermons and lectures not to speak of letters and editorials, will undergo a revolution as remarkable as that effected in books by the invention of printing and the weary process of learning penmanship in schools will be reduced to the acquirement of the art of writing one's own signature, and playing on the literary piano above described, or, rather, on its improved successors."

The history of "typewriting" is a long and involved one. Many people had "invented" machines, but by 1867 there was no practical device in use. The first recorded device was known because Queen Anne of England had granted a patent to Engineer Henry Mill in 1714 for . . . "An Artificial machine or Method for the Impressing or Transcribing Letters, Singly or Progressively one after another as in Writing, Whereby all Writing whatever may be Engrossed in Paper or Parchment so Neat and Exact as not to be distinguished from Print; that the said machine or method may be of great vse in settlements and publick recors, the impression being deeper and more lasting than any other writing, and not to be erased or counterfeited without manifest discovery; and having, therefore, humbly prayed vs to grant him our Royall Letters Patents for the sole vse of his said Invention for the term of fourteen yeares, etc. . . . ." But it never came to be.

The first recorded effort to build a writing machine in America was that of William Austin Burt of Detroit, Michigan, to whom, on July 23, 1829, the U.S. Patent Office granted a patent "for a machine for printing" which he self-styled the "typographer". "Burt's Family Letter Press" was never manufactured but it was talked about. In May, 1829 a correspondent for the *New York Commercial Advertiser* described it as "a simple, cheap, and pretty machine for printing letters." And the editor commented: "It should be found to fully answer the description of it." Both editor and correspondent could not suggest a name for it, a point on which Burt had solicited advice.

Our gang of backyard inventors spent more time talking than working. Inspired by what they had read in *Scientific American* about John Pratt's writing machine, Glidden noted that the paging device was something of a writing machine and repeated: "Why can't you make one that will print words as well as figures?" "I can," Sholes replied. "I've thought about it a great deal, and I'm going to try it."

Sholes did think of something: he cut a type character on the side of a short bar made to strike up against a piece of paper held beneath a round glass disc, almost like a pianoforte's hammer striking from below against the strings. He told Roby who promptly started to whittle one of the parts out of wood. The whole gang decided the idea was worth a shot.

The first model was was simply the letter "W" and its activating mechanism, a telegraph key linked to a type bar pivoted to hit from below against a round glass disc. It was demonstratable in July at the expense of the magic clock, the pager and the spading machine. Sholes got some carbon paper from the Western Union office, the only place in Milwaukee that had some of this unique material.

Sholes held a sheet of paper and the carbon together. With his left hand against the underside of the glass he drew them slowly from right to left tapping the telegraph key with his right hand.

The result was a line of Ws.

By September he finished another working model, with black walnut, piano-like keys lettered in white. At a small party at Kleinsteuber's Sholes sat down at the machine and typed:

"C. LATHAM SHOLES, SEPT. 1867."

It was in all caps because there was no lowercase.

Sholes had learned a lot in building the first models and thought that he was near completion; he did not know that he would be building machine after machine for the next six years. He did have a name for the machine, which was still called "the machine". The others in the group wanted a name that was distinctive, like "writing machine" and "printing machine."

Sholes thought of "type-writer"—with the hyphen—based on his experience as a printer.

The next model was less than perfect, but before Sholes took it apart he typed the first letters ever typed to people who might be able to help him, offering shares in return for funds. He wrote to many people whom he thought might be interested. One reached James Densmore of Meadville, Pennsylvania, a lawyer, promoter, salesman and inventor himself, whom Sholes had met in Madison some twenty three years earlier. Densmore had been a newspaperman and job printer at the time, and Sholes remembered him as a giant of a man with shaggy red beard. It was said that Densmore had made a killing in the Pennsylvania oil boom by inventing and patenting the first oil tank car.

Sholes received a reply from Densmore saying he wanted a part in the writing-machine. The terms were that Densmore would pay the group's back bills, about $600 (although some reports state that the amount was $6,000), and provide future financing in return for a 25% interest. Sholes did not know that $600 was Densmore's total liquid assets. Whether he was loaded or broke, Densmore looked terrible—with seedy, battered hats, shabby coat and trousers inches too short. He always appeared to be on the verge of penury.

The first machine was finished by 1867. Patented with improvements June 23, 1868, it became patent number 79,265 and was listed as "Sholes et al." It had eleven piano keys and was of wood. Historians do not usually refer to the machine of June 23, 1868 patent, but speak of the "first machine" as the July 14, 1868 patent since it was the first working model sent out from Sholes' shop for testing by someone other than himself.

The July 14, 1868 machine had a circular disc with radial grooves and slots to receive and guide the typebars so they struck the center, radial typebars to correspond with this disc; a ratchet to move the paper carriage by the width of a tooth when a key was struck, and a hinged clamp to hold the paper in its carriage. Most of Sholes' machines did not show the type as it was being typed. A visible machine was made later by having the type strike the paper first and getting its impression from the inked ribbon passing underneath. The later introduction of the roller brought with it loss of the visible feature but made possible use of paper of any thickness, instead of tissue paper.

The typebars were straight brass rods with letters cut on the ends that were apt to bunch together in the slots at the striking center. When a ribbon wore out you bought a bolt of silk or satin cloth, unrolled and dipped it into black ink, then strung it out to dry.

Densmore, or someone like Densmore, was an important factor in the development of the device, and his name is often linked with that of Sholes. Without him, the Type-Writer might never have been. He provided optimism and money, two very important contributions. He invested in the machine before he had even seen it, becoming its loudest and most fervent champion. In the darkest hours when Sholes was ready to give up, he urged the would-be inventor onward and gave hope.

Densmore argued that the Type-Writer needed work and he forced Sholes through nearly fifty different models. It must be emphasized that Sholes was a tinkerer, not an inventor and much of the work resulted from trial and error. Densmore turned the experimental machines over to professional stenographers to give them the toughest tests they could possibly devise. And they found the problems: the type bars tended to stick and bunch; the weight on the clockwork device that moved the carriage was too light and its string was liable to break; it stuttered and jumped; while the hand-inked ribbon was a mess.

We now meet James Ogilvie Clephane. He was born in Washington, D.C., February 21, 1842, of Scotch parentage, studied shorthand at an early age and was eventually considered the best in his chosen profession. At the important trials taking place in the National Capital, he was the stenographer of choice. Dissatisfaction with reproduction approaches led him to review alternatives.

Clephane was a short man, given to wearing high-heeled shoes and tall hats to give the effect of height. It was said that he was a "Napoleon" of a man who could not accept defeat. He was admitted to the Bar of the Supreme Court of the District of Columbia and had several offers to enter the practice of law, but his interest in inventions became so great that he devoted his entire life to them, especially as they related to the printing processes.

Clephane and John H. White, official reporter of the House of Representatives, and Andrew Devine who had the same post in the Senate, wanted to find some way of transcribing reports faster and then making multiple copies. The Type-Writer was the first device to excite their imaginations.

It was natural that Densmore would have wanted Clephane to review the Type-Writer and equally natural that Clephane would want to do it. At the same time Clephane was urging forward other inventions in this line and expending large amounts of capital for the purpose.

Other stenographers attacked defects in Christopher's models, about thirty of them, each with some change, usually intended to reduce friction and heighten speed. Stenographers were expected to be among the first to use Type-Writers and Clephane rigorously tested all models The five or six years' work on these models has been termed the "most remarkable part" of Sholes' whole effort. Clephane was so unsparing in his tests that he often reduced a machine to ruin. His judgments were so caustic that even Sholes lost his temper, exclaiming, "I am through with Clephane!"

Clephane's frank reports irritated and disheartened the inventor, and Densmore would persuade him to keep making improvements. "This candid fault-finding is just what we need," said Densmore repeatedly. "We had better have it now than after we begin manufacturing. Where Clephane points out a weak lever or rod, let us make it strong. Where a spacer or an inker works stiffly let us make it work smoothly. Then depend upon Clephane for all the praise we deserve."

Later, in the fall of 1874, the typewriter would be used in a celebrated burglary case. Clephane used seven Remingtons and had seven stenographers read their notes to the typists. Using carbon paper, he was able to have three sets of transcripts revised, stitched and indexed by eight o'clock each evening, for six weeks.

Clephane also was interested in the Phelps Printing Telegraph, used by Western Union, but felt that adapting it to printing would primarily benefit Western Union. He found another printing telegraph at the Patent Office, the invention of Charles T. Moore, a young West Virginian. He contacted Moore and began to develop the device into a reproducing printer but the project was not successful.

In 1875 Clephane continued his interest in a typewriting machine which had been evolved by George H. Morgan, of

Ulrichsville, Ohio. Clephane at this stage was assisted by Peter J. Hannay and Andrew McCallum, who were Patent Attorneys in Washington. Clephane and his friends, especially White and Devine, were not great capitalists, but they put their savings into a company called "The Typographic Machine Printing Company" and proceeded to try to develop the idea of mechanical writing and typesetting.

Then returned the persistent Mr. Moore who showed Clephane a machine he had devised for typing on a typewriter with lithographic ink impregnated on a paper ribbon. We will interrupt this story line and continue with Sholes, but Clephane and Moore will be back.

Sholes continued to improve his models in the light of Washington objections. In all, fifty or so were tried before Densmore was satisfied. Each cost about $250 and each possessed one or more improvements over the last.

One of the first test lines was the current political slogan, telling Republicans to forget their internal squabbles and concentrate on helping Ulysses S. Grant beat Horatio Seymour: "Now is the time for all good men to come to the aid of their party."

On July 30, 1870, Sholes wrote to a friend about the latest version of the Type-Writer:

. . . I think the machine is now as perfect in its mechanism as I know how to make it, or to have it made . . . The machine is done, and I want some more worlds to conquer. Life will be most flat, stale and unprofitable without something to invent.

As with Hamlet, Sholes perfect mechanism was not to be. Five months later he was working sixteen hours a day on an improved model. Kleinsteuber's shop was not properly equipped and Sholes jumped from shop to shop, from West Milwaukee to rented premises in Chicago, where fifteen Type-Writers were turned out. The landlord wanted his money before the machines were sold, and the inventors returned to Milwaukee.

Few Type-Writers were sold because most of the production was either given away for publicity purposes, or undergoing tests, or held by creditors. Everyone was discouraged except Densmore. The expenses of maintaining his optimistic facade forced him to live on raw apples and soda crackers. But he was pushing ahead, at one point he telegraphed Sholes from New York that he had a deal with giant Western Union and he needed Sholes, in person. They called on the president of the company, with lanky and frail Sholes carrying the bulky Type-Writer and beefy, unsartorial Densmore leading the way.

Western Union was interested, but wanted some of its telegraphers take a look at the machine. Time passed and no word was forthcoming. Densmore was not able to stand the suspense. He went back and asked. The explanation was that they liked the machine, but one of their employees, Thomas Alva Edison, thought he could do better for less than the $50,000 Densmore wanted.

Early in the seventies Sholes, accompanied by a friend named Craig, went to Tom Edison's shop in Newark for help and TAE was able to give him some assistance. Edison did some typewriting inventing of his own. He made the first practical metal typewriter and his patent of December 10, 1872 for an electrically operated traveling wheel device (Patent number 133,841) was the forerunner of the stock ticker printing machine. Edison found use for typewriters in automatic telegraphy.

Densmore was still optimistic. "I believe in the invention from the topmost corner of my hat to the bottom-most head of the nails of my boot heels," he declared. His enthusiasm and his share was gradually increasing. As funds were required for patent models on each improvement, or to pay urgent bills, Densmore argued that he had to get something for raising them. One way he could be compensated was by increasing his interest in the invention. The original shareholders, with no profits in sight, turned their shares over to Densmore, until, by 1872, Glidden, Soulé, and Roby of the original group had nothing left. Sholes still owned some share, but not much.

A new participant, George Washington Yost, purchased part of Densmore's own interest and then persuaded Sholes to sell his rights for a flat fee. That amount is disputed and may have been anything from $1,200 to $40,000, but probably $6,000.

Yost was said to be a smooth talker. He was a New York State farm boy but he talked well enough to be mistaken for a man of social distinction. The 1873 Sholes-Glidden-Soulé machine, which Densmore and Yost now owned, had only faint resemblance to the first Milwaukee model. It had a four-bank keyboard with the letters and numerals arranged in what we now

Christopher Latham Sholes, principal inventor of the typewriter.

Sholes' daughter was among the first operators of the typewriter.

The article on the "Pterotype" provided the inspiration to Sholes and his friends.

The single letter (W) key that Sholes used to test the idea of the typewriter.

The Burt "Typographer".

First working model of the typewriter (June 23, 1868).

An interesting illustration of the typewriter that Mark Twain supposedly purchased. Note the lower right-hand corner: "Clephane & Co. Artographers & Lithographers"

Ottmar Mergenthaler

James Olgilvie Clephane was one of the links between the typewriter and the Linotype.

Second working model of the typewriter (July 14, 1868).

A Remington Model No. 1 without stand (1873).

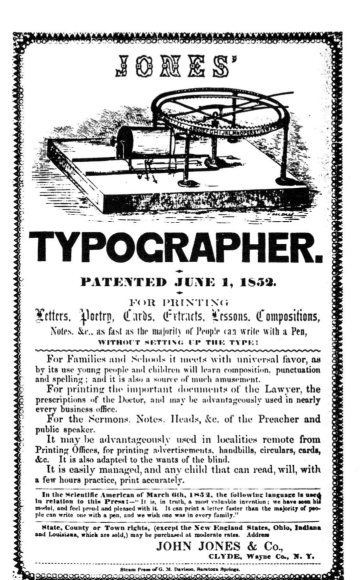

An ad for the "Typographer" which carried a quote: "It can print a letter faster than the majority of people can write one with a pen".

Ottmar Mergenthaler, fourth from the left, stands with his early employees and the first set of Linotypes in his Baltimore factory.

The first three of Ottmar's five children.

Ottmar and his wife, Emma.

call "standard" order. The machine was partially encased in a japanned tin case, squarish and tall. The carriage was moved by the weight of a leaden ball on a line that hung down the right-hand side of the machine. The major difference between the 1873 concept and modern typewriters was the understrike—the type bars hit upwards against the bottom of the roller. The printing point was out of sight; the operator could only see what had been typed three or four lines after the fact.

The two new partners found their way to Philo Remington, eldest of Eliphalet Remington's sons, and president of the family business—making guns, sewing machines and farm machinery. The Remingtons had a factory at Ilion, New York. Densmore and Yost took a room at Small's Hotel, set up the Type-Writer, and demonstrated it to Remington and others. Yost did the talking. A contract was signed on March 1, 1873. Remington agreed to dedicate an entire wing of the plant for Type-Writer production and to manufacture one thousand of the machines. Densmore and Yost were to be the selling agents for the product.

Remington put two mechanics on the job, William K. Jenne, who had been head of the sewing-machine division and would continue with the unit for the next thirty years, and Jefferson M. Clough, who had the task of remodeling the machine for quantity production. Jenne and Clough solved many of the problems that had haunted the original team. Manufacture began in September and first shipments were made early in 1874.

Jenne liked sewing machines and as a result the first Remington No. 1 Type-Writer looked like a sewing machine, with a foot-treadle carriage return and flowers stenciled on its sides. The public did not seem to really care, considering the outrageous price. There were some takers and Mark Twain was one—we shall see him again among those interested in typesetting.

On a lecture tour in Boston, Twain was taking a stroll with humorist Petroleum V. Nasby (whose real name was D.R. Locke) when they saw a Remington Model 1 on display. A salesman claimed that it could "write" at a speed of fifty-seven words a minute. Twain did not believe it could. The salesman called a young lady and Twain and Nasby timed her. She typed fifty-seven words in almost sixty seconds. Twain said that it was luck and she did it again.

Twain put down his $125 and bought a Type-Writer to be

delivered to his home in Hartford, Connecticut. Twain says, in his autobiography:

"We got out our slips and were a little disappointed to find that they all contained the same words. The girl had economized time and labor by memorizing a formula which she knew by heart.

At home I played with the toy, repeating and repeating and repeating "The boy stood on the burning deck" until I could turn out that boy's adventure at the rate of twelve words a minute; then I resumed the pen for business, and only worked the machine to astonish inquisitive visitors. They carried off reams of the boy and his burning deck."

When he got back to Hartford, he wrote a letter on December 9, 1874, to his brother, Orion Clemens:

DEAR BROTHER:
I AM TRYING TO GET THE HANG OF THIS NEW FANGLED WRITING MACHINE, BUT I AM NOT MAKING A SHINING SUCCESS OF IT. HOWEVER THIS IS THE FIRST ATTEMPT I HAVE EVER MADE & YET I PERCEIVE I SHALL SOON & EASILY ACQUIRE A FINE FACILITY IN ITS USE . . . . THE MACHINE HAS SEVERAL VIRTUES. I BELIEVE IT WILL PRINT FASTER THAN I CAN WRITE. ONE MAY LEAN BACK IN HIS CHAIR & WORK ON IT. IT PILES AN AWFUL STACK OF WORDS ON ONE PAGE. IT DON'T MUSS THINGS OR SCATTER INK BLOTS AROUND. OF COURSE IT SAVES PAPER . . . WORKING THE TYPE-WRITER REMINDS ME OF OLD ROBERT BUCHANAN, WHO, YOU REMEMBER, USED TO SET UP ARTICLES AT THE CASE WITHOUT PREVIOUSLY PUTTING THEM IN THE FORM OF MANUSCRIPT. I WAS LOST IN ADMIRATION OF SUCH MARVELOUS INTELLECTUAL CAPACITY . . .

YOUR BROTHER,
SAM

In March, 1875, Twain wrote to Remington:

Gentlemen: Please do not use my name in any way. Please do not even divulge the fact that I own a machine. I have entirely stopped using the Type-Writer, for the reason that I never could write a letter with it to anybody without receiving a request by return mail that I would not only describe the machine but state what progress I had made in the use of it, etc., etc. I don't like to write letters, and so I don't want people to know that I own this curiosity breeding little joker.

Yours truly,
Sam L. Clemens.

"Tom Sawyer" was published in 1876 and Twain's autobiography says that the manuscript was typed before it was submitted. The Herkimer County (New York) Historical Society believes that Twain's memory confused "Tom Sawyer" with "Life on the Mississippi" (1883). Twain was the first author, it is said, to submit a typewritten book manuscript, with a double-spaced, single-side format that has thrilled editors ever since.

It is said, probably in jest, that Twain once submitted a manuscript with no punctuation. The editor wrote back and pointed this omission out, whereupon Twain sent a sheet filled with periods, commas, etc. with instructions to place them in the manuscript accordingly.

Although Nasby did not buy a Type-Writer he gave up the lecture racket and started selling Type-Writers himself, becoming a partner in the organization that bore the name, for a time, of Densmore, Yost & Locke.

Remington exhibited at the Philadelphia Centennial Exposition in 1876. For 25 cents, a comely lass at a Type-Writer would type a brief note suitable for mailing to the folks back home. Remington sales still did not take off. The exposition was a disappointment to Sholes, who expected notoriety, but another inventor named Bell got much of the attention.

Remington sales rose slightly in 1877 when an Ithaca, New York, sales agent, named William Ozmun Wyckoff, started a periodical called "The Typewriter Magazine", which, according to its masthead, was "devoted to true reform, to the welfare of all mankind, and to advancement and improvement in all things." The Typewriter Magazine did start to get people interested in the Type-Writer.

Yost then visited Erastus Fairbanks, inventor of the Fairbanks scales, at St. Johnsbury, Vermont, who was selling weighing machines. Yost persuaded Fairbanks to take over sales of writing machines and Fairbanks put 23-year-old, Clarence Walker Seamans, in charge of the Type-Writer. The job of selling Type-Writers proved too tough for the Fairbanks outfit and sales went back to E. Remington & Sons, which had been manufacturer only. Remington took Seamans away from scales and kept him in Type-Writers. Sales in 1881 came to twelve hundred machines.

Seamans and Wyckoff went to Henry H. Benedict, a Remington executive and one of those who supported the Type-Writer. The three of them formed a new company on August 1, 1882 and bought the Remington Type-Writer sales rights in the United States (and the entire world for that matter). They committed to take as many typewriters as Jenne and his factory could manufacture. Wyckoff, Seamans & Benedict opened offices on lower Broadway in Manhattan and within five years owned the Remington Type-Writer, including the wing of the Ilion factory, all the patents . . . the works. From March, 1886 on, Remington Type-Writer and E. Remington & Sons were separate companies.

Back in 1878 the problem of printing both capital and small letters was solved by placing large and small type on a single key and shifting them with a shift-key mechanism. This Remington model used the cylinder shifting device of Lucien Stephen Crandall, and the double types, a capital and lowercase of the same letter mounted on the one type bar, invented by Byron A. Brooks. Remington's patent attorney, Phillip Tell Dodge, had secured the patent; the company criticised him for acting without authority, but finally realized the importance of the two-letter key.

A machine with forty keys, two being shift keys, could print seventy-six characters with both capital and small letters. The machine was the No. 2 Remington.

Sholes typebars had a tendency to collide and stick to one another at the printing point because of their arrangement. So he and Glidden arranged the keyboard in such a way that the letters which occurred most frequently in the English language were placed as far as possible apart in the typebasket. Sholes, Glidden, and Soulé had their inventive limits and never thought that anyone would want to use more than two fingers. The three patterned the keyboard layout roughly after the distribution of letters in a printer's case for handset type, and the pattern stuck. Although there are folks who believe that the key layout was based on the fact that Sholes was left-handed, we may never know.

Franz X. Wagner is credited with the first visible sold to the public. John T. Underwood bought it, gave it his name, started manufacture 1894-95. Christopher's last patent was granted August 27, 1878. James B. Hammond, inventor of the typewheel machine which bears his name, felt in 1880 that the arrangement of keys on the early Remington could be improved. Accordingly, he devised the "Ideal" keyboard. It was supposed to give the operator much greater speed than could be obtained from the "American Standard" because the characters most frequently used were put in such positions that the fingers of the operators could manipulate the keys faster. The Hammond typewriter evolved into the Varityper—a typewriter typesetter and the first "cold type" or strike-on device.

Typewriting was the opportunity that opened the door for the emancipation of women. From typing it was a natural step to general secretarial and office work, and then into business and industry. Sholes was always glad he was the one who helped womanhood by providing them with an opportunity to enter the business world:

"I don't know about the world, but I do feel," he said shortly before his death, "that I have done something for the women who have always had to work so hard. This will enable them more easily to earn a living."

"Whatever I may have felt in the early days of the value of the typewriter," he wrote in one of his last letters, "it is obviously a blessing to mankind, and especially to womankind. I am glad I had something to do with it. I built wiser than I knew, and the world has the benefit of it." Thus wrote one of the great, but somewhat ungrammatical, unsung inventors of the 19th century.

Sholes died of tuberculosis in Milwaukee on February 17, 1890 at seventy one.

He is barely remembered and never honored. A park in Milwaukee was named in his memory, and then because few could remember why, the park's name was changed. No device or company carried his name.

Christopher Latham Sholes tinkered his way into a footnote to history. For over 100 years, his device was the basic tool for written communication. Word processing and electronic text editing have yet to displace the typewriter.

CHAPTER 2

THE DAY THE CLOCK'S HANDS MOVED

In June, 1867, in the village of Ensingen, Germany, the hands of the clock on the old Lutheran Church moved. Ottmar, son of schoolmaster Johann Mergenthaler had fixed the clock, even though a clockmaker said it could not be repaired. Ottmar told his story for the benefit of his family, explaining how he had gotten the idea by studying the illustrations in a book on clocks sent to him by his stepmother's brother, Louis Hahl, a watchmaker in the town of Bietigheim.

He had managed to sneak up to the deserted tower every afternoon while the pastor was out of town. He had taken the entire clock mechanism apart and discovered a broken pin which he replaced, scrubbing the rust and grime from the gears and springs before putting the works together.

His father, Johann Georg Mergenthaler had come from a family of peasants, living in the kingdom of Wurttemberg in southwest Germany. With a stubborness that would characterize his son Ottmar, Johann refused to accept the peasant life and enrolled in a training school for teachers. In 1849 he married Rosine Ackermann, and in 1851 she gave birth to a son, Gustav. The following year another son, Karl, was born. The Mergenthalers moved to the town of Hachtel and on May 11, 1854 a son named Ottmar was born. The next year a fourth child came into the world, a daughter named Caroline.

In 1859 Rosine developed pneumonia and died. Two years after Rosine's death, Johann married Caroline Hahl, who took the children as if they were her own. Caroline came from a family of merchants and artisans and was gifted with common sense. "It is not important who you are but what you do," she instructed the Mergenthaler children.

It was Caroline who recognized Ottmar's mechanical ability and encouraged him to develop it. He was so handy that she nicknamed him "Little Cleverhead."

Caroline asked her brother Louis, a watchmaker in Bietigheim, to save his unneeded tools and metal odds and ends for her stepson. Schoolmaster Mergenthaler objected to Ottmar's preoccupation with tools and mechanics since it had been his dream to have his sons become schoolmasters. Without education he would still be toiling in the fields like his ancestors and he constantly impressed his sons with the need to study.

Johann could not afford the tuition and board which university attendance demanded and knew that if his children were to achieve a decent education, it would have to be at seminaries. Ottmar resisted this idea and announced that he wanted to become an engineer.

Johann pointed out that engineering meant attendance at an expensive technical institute which was also out of the question financially. Ottmar was defiant, stating that if he could not be an engineer, then he would become a mechanic.

In 1868 Caroline suggested that they apprentice the boy to her brother. Louis Hahl, his back hunched from a lifetime of leaning over work benches, was precise and methodical. A widower, he had a son who had learned the watchmaking craft and then emigrated to the United States to start a shop of his own. He felt a deep affection for Ottmar and recognized the boy's gift for mechanics. Therefore, when Caroline suggested that he take on the youngster as an apprentice, Louis Hahl accepted the idea, in spite of the fact that six apprentices and journeymen were already employed in his shop.

The agreement between Johann Mergenthaler and his wife's brother called for Ottmar to serve for four years without wages. In return, Ottmar was to receive training, lodging and occasional holidays to visit his parents. The new apprentice took to watchmaking with a skill that astonished everyone.

Ottmar found in machines a joy that others saw in art. Ottmar was to look back on his early training and recall: "Above all watchmaking taught me precision. I learned to temper a spring to the finest degree, to combine the constituents of metal alloys in exact proportion. I learned how to cut the finest teeth, to make pins, to bore jewels with firm, steady pressure. I realized that if a movement was to work it must be considered as a whole, that each part had to be perfect in itself and also harmonize with every other."

Special schools to enable young workingmen to better their education were established with free classes held in the evenings and on Sundays so that the students would not have to lose time from their jobs. Ottmar registered for courses in mechanical drawing and blueprint reading. He felt it would be helpful to learn how to present mechanical ideas on paper. He also took a course on electricity, which the miracle of telegraphy required.

Germany had been a land of divisive conflicts. The nation was under the control of Count Otto von Bismarck. In less than a decade, Bismarck had expanded the army, engaged in a number of wars and merged the German states into a strong Confederation. Germany had recently triumphed over France in the Franco-Prussian War of 1870. The excesses of the new central government stirred up considerable opposition and bitterness, particularly in the south.

In 1871, when Ottmar was in his third year of apprenticeship, his two older brothers were drafted into the army. Germany was constantly on the brink of war.

The apprenticeship agreement called for Ottmar to serve for four years without pay. However, he had progressed so rapidly that at the end of the third year Louis announced that he would receive journeyman's wages.

In the spring of 1872 Ottmar explained that he was thinking of leaving Germany. He had read the encouraging letters Uncle Louis had received from his son in the United States. August, now thirty, had emigrated to America and had opened a shop for the manufacture of precision instruments in Washington, DC and was prospering. August wrote of the opportunities for European-trained craftsmen in the bustling nation. Louis learned that Ottmar's parents had given their permission for him to go to the United States; he then wrote to his son and Ottmar received a job offer. "If Ottmar has developed into as good a craftsman as you say, I would like him to work for me," August wrote. "I will even advance him the money for his fare, which he can repay out of wages."

In October, 1872, Ottmar, accompanied by his family, went to Stuttgart to board the train for Bremen, the great seaport in northern Germany. The ship "Berlin" was the pride of Germany's merchant fleet, but Ottmar and almost five hundred other emigrants were herded into steerage and saw little of its grandeur. The crossing took twelve days, landing in Baltimore on October 26, 1872.

He took the train to Washington, DC where he met his stepcousin, August, who had come to the United States in 1864 while the Civil War was still raging. There had been a demand for skilled craftsman and he had easily managed to get a job in a Washington factory and had accumulated enough money to open a watchmaking shop after the war.

From watches and clocks he expanded into the manufacture of other precision instruments. His reputation for reliability brought him to the attention of government officials, and he built devices for the U.S. Weather Bureau, including heliographs, wind velocity registers and rain gauges.

August's location in the nation's capital led to the building of models of new inventions. Every inventor applying for a patent was required to submit a model of his invention to the Patent Office. The Patent Building, which had been used as a hospital during the Civil War, exhibited the models in the grand hall. The number of inventors was staggering; it was truly an age of invention. They came to Washington with blueprints, sketches and ideas by the thousands, many without the vaguest notion of how to build models of their brainchildren. Model building became an industry in Washington, and August Hahl drew a great deal of this business.

Ottmar found himself in a world of organized bedlam. August employed a dozen workmen and demonstrated a genius for supervising them while handling the multitude of inventors who found their way to his shop. If an inventor came in with an idea that was unsound, August explained why it was impractical. If the inventor insisted on having the model built, the job would be accepted at a fair price, with the understanding that the shop could not assume responsibility for its success. If an idea appeared sound, but the inventor did not have money, August would construct the model anyway, on a royalty basis, although these subsidy arrangements had not proved profitable.

Ottmar was assigned to work repairing watches and clocks. He was quickly mastering English and had so impressed his stepcousin with his skill and efficiency that he was transferred to more important work. His English improved so rapidly that it was hard to believe he had been in the country only six months.

August assigned him to work on experimental meters and electrical apparatus for the U.S. Signal Service, a branch of the Army. It was a contract that called for building equipment from blueprints and improving the design where possible. Ottmar's mechanical drawing and electricity courses made him one of the few men in the shop who had any real ability to handle this contract effectively.

He was working closely with the U.S. Signal Service, including the engineers who had drawn up the original designs. Ottmar discovered that inventors were a unique breed and he learned from each of them.

By 1873, the country was in a financial panic as business came to a virtual standstill and bankruptcy coupled with growing unemployment began to take shape. Contracts were canceled and it was apparent that things were not improving at all. August informed his wife Gerda and Ottmar that he would move the shop to Baltimore, a more important industrial center than Washington, where there was a greater likelihood of contracts.

In January, 1874 August sold much of his machinery at a loss and moved the remaining equipment to 13 Mercer Street, in the heart of Baltimore's industrial section. Ottmar continued to board with them. Business was as bad in Baltimore as in Washington, but with lower overhead he was able to survive.

Ottmar refused to accept wages and told August to deduct any amounts from what he still owed for his passage. Just when their situation appeared to be at its worst they received a contract for signal devices to be used at the Philadelphia Centennial Exhibition which was to take place in the summer of 1876. Soon the machines in the Hahl shop came alive again. August hired additional workers and, Ottmar, who had just turned twenty, was foreman.

Business was picking up at last as they received visits from a few inventors who had been referred to the Hahl shop by government officials in Washington.

On August 17, 1876 a tall, bearded stranger entered the shop carrying a bundle of sketches. His name was Charles T. Moore and he had just come up from Washington, he said in his slight southern drawl. The two cousins did not know it but Charles T. Moore and the diminutive James Clephane were to be the spark that ignited the genius of Ottmar Mergenthaler.

CHAPTER 3

THE TRANSFER TYPEWRITER

Moore had tried his hand at inventing a typesetting machine. In 1872, his Double-Cylinder Typesetter was patented, but considered impractical. This hot afternoon, he unrolled his blueprints and sketches for a table-top machine composed of rows of finger keys and a cylindrical roller containing successive circles of metal letters of the alphabet. Moore, who lived in White Sulphur Springs, West Virginia, explained that by tapping the keys as the cylinder revolved, the metal characters would be imprinted in lithographic ink on a thin ribbon of paper. This strip would then be cut into segments the width of a newspaper or magazine column and transferred to a lithographic stone for printing. The strips could be snipped at any point spaces between words and the space increased or decreased in order to justify the columns of type.

He called the machine a "transfer typewriter" and said that if they could build a model that worked, his backers were willing to pay $1,600.

He did not tell them that a great deal of time and money had already been expended in trying to get the process to work. A month before, on July 3rd, a demonstration in Baltimore had not gone well. Moore and Clephane had then gone down to the Centennial, all two hundred buildings of it, to learn what they could about new developments in printing. They returned with optimism that their invention would revolutionize the printing industry. A demonstration set for July 6th resulted in an altercation between Moore and his machinist. And so it was decided to find another machine shop.

Ottmar criticised the basic design and Moore was momentarily offended at these comments but continued. "If you are willing to take on the job, I'd like to arrange a meeting with my backers at the earliest possible moment. You may have heard of one of them—James O. Clephane."

It was said that everyone in Washington knew of Clephane. He was the best known shorthand reporter in the capital. During the Civil War he had been private secretary to Secretary of State William H. Seward and a friend of Presidents James Buchanan and Abraham Lincoln. His speed and accuracy in taking shorthand notes was so renowned that he was called in by the government to record the most important congressional and legal proceedings.

Clephane had foreseen the tremendous significance of mechanical devices for his own field of shorthand reporting and had worked on the invention of "Type-Writer." Clephane had been responsible for attracting public attention to Sholes' invention by using seven machines and operators to transcribe shorthand notes during a celebrated criminal trial. It was well known that he was very interested in the printing process and had supported the development of several approaches. He was called "the godfather of inventions".

Neither August nor Ottmar had seen a typewriter, but when they learned of Clephane's interest in Moore's "transfer typewriter" they recognized points of similarity between it and the Sholes' machine, which had been described in many newspaper accounts. Both used finger keys and metal characters to print directly on paper. August commented on the resemblance and Charles T. Moore admitted that Clephane had formulated the principle after seeing the typewriter.

The full name of one of Clephane's companies was "Clephane & Company—Artographers and Lithographers". One of the major tasks that they faced with shorthand copy was its reproduction. The artographic press was the method that Jefferson and the founding fathers used to make copies of their correspondence. (The model of the double pen apparatus shown in Monticello was not used very often.) Colonial communicators wrote with very liquid inks that were transferred by pressure in a desktop press to tissue paper. The images were reversed but then read through the back side of the light-weight papers.

Lithography had been discovered in 1791 by Alois Senefelder, a Bavarian composer. He was attempting to find a method for reproducing his music compositions when his experiments led him to what he called "stone printing" or "chemical printing." He used a polished block of limestone on which he wrote with an ink or crayon compound. After the stone was washed with water, ink was applied and adhered only to the image areas. The image had to be in reverse form so that transfer to the paper brought it back to right-reading form. Lithography means "stone writing" in Greek.

August agreed to discuss the project further and went to Washington to see Clephane and the other backers. When he returned he informed his stepcousin that he had accepted the contract for sixteen hundred dollars with payment only if the Moore machine was made to work. In other words: no results, no payment. Clephane had agreed to send the earlier model to the Hahl shop for analysis, in the hope that it would help them find the reason for failure.

August had met Clephane himself as well as his brother Lewis Clephane and two businessmen, Maurice Pechin and J. H. Crossman. All of them had fallen under Clephane's spell and were convinced that Moore's machine was destined to revolutionize printing.

The original model arrived at the Hahl shop and August turned it over to Ottmar for study. It had been made in a Baltimore shop owned by a man named Grant. In spite of Moore's complaint, the quality of workmanship was not faulty; the parts had been machined out of the best materials available and for the most part the mechanism had been assembled with reasonable care. August asked whether it could be improved. Ottmar declared that he could, but added ". . . even though I know almost nothing about printing, I have little faith that this is the machine to revolutionize an industry."

A few doors away from the Hahl shop on Mercer Street was a printing establishment owned by a Charles W. Schneidereith. His son Louis was about the same age as Ottmar and the two had become friends. Ottmar would sometimes stop in to chat with Louis and would watch the compositors pick up their "sticks" and assemble lines character by character. But he still had no real understanding of the printing process.

The Moore machine was designed to utilize the lithographic printing process, a process based on the fact that water and greasy ink repel each other. Lithographers also found that they could take an image created on paper or sheets of metal with lithographic ink and "transfer" it to stone for reproduction. It was this "transfer" principle that the Moore machine sought to incorporate. By printing characters on a strip of paper with lithograhic ink and transferring them to a polished stone, it was hoped that one could key characters for typesetting rather than the laborious task of assembling handset type or drawing type directly on the litho stone.

Finally, in October Clephane came to Baltimore. He was short, red-faced and portly, with an impressive beard, hardly appearing to be a promoter who was able to marshal inventors and financial backers to his cause. His voice was filled with self-assurance.

Ottmar tried to explain that lithography called for skillful, patient craftsmen and would not be a practical process, Clephane ignored this point: "I am a man possessed by an idea. I intend to overcome all obstacles to see it carried through. If you tell me that it cannot be done, I will answer in the words of Napoleon who, when he wanted to cross the Alps and was told of the difficulties, cried, 'There are no Alps.' I tell you now: the machine has got to be built."

Clephane brushed aside all objections and went into a detailed explanation of the reasons why he was so concerned with the success of the Moore machine. As a shorthand reporter he had long felt the need for a quicker method of transcribing notes. The Remington typewriter invented by Christopher Sholes had solved a part of this problem. Now he wanted a machine that would make a great many copies of a given manuscript. "I want to bridge the gap between the typewriter and the printed page," he said prophetically.

Ottmar went to work improving the Moore invention. The design of the cylinder had placed both roman and italic characters on a single unit. Ottmar re-designed it as two units, one for the roman and one for the italic characters. He prepared a new set of drawings, using the double-cylinder principle and incorporated a shift key, like that on the latest Remington typewriter.

Ottmar found that parts could be eliminated and the basic

structure of the machine simplified. By January, 1877, he had completed a small working model. August sent Clephane a telegraph message informing him of the progress and requesting further instructions. Clephane rushed up to Baltimore in person, examined the machine anxiously and observed that it differed considerably from Moore's original design.

Ottmar threaded in a strip of paper and began to manipulate the keys. He printed his name several times, snipped off the strip of paper and handed it to Clephane.

"By heavens, this is remarkably clear." Clephane exclaimed. He typed several sentences himself, examined the paper ribbon and announced that he was satisfied. Ottmar ventured the suggestion that they test several of the printed strips in a lithography shop to see if faithful copies could be produced, but Clephane had to return to Washington at once. "I'm certain," he added, "that with print as clear as this we will have no problem." Before leaving, he authorized August and Ottmar to proceed with construction of a full-sized machine.

All during the spring months Ottmar worked at this task with scarcely a day passing without a letter or telegraph message from Clephane inquiring about his progress. By July the full-sized machine was ready. Clephane wired back that he had arranged to have tests made at a lithography shop in Washington and instructed them to bring the machine along. In the Capital they found the four backers present as well as inventor Charles T. Moore, who had come over from West Virginia. The machine was taken out of its crate and set up on a table. Ottmar threaded in the paper strip and tapped the finger keys. Between the keys and the paper was a cloth ribbon inpregnated with greasy lithography ink. As each key struck, it left an inked impression on the paper. After typing several words, Ottmar tore off the paper strip and passed it around for the others to examine. Everyone expressed pleasant amazement at the surprising clarity of the letters, exclaiming "Why, it's better than anything expected."

The paper strip was handed to the lithographer, who pressed it against the flat polished stone, transferring the greasy characters to the smooth surface. The stone was washed down with water and a roller used to apply a thin layer of printing ink to the line of characters. The process was explained. "The ink is water-moistened, repels the ink. Now see what happens." The lithographer applied a sheet of paper to the surface and used a clean roller to obtain a steady, even pressure to transfer the image. He peeled off the copy, studied it and handed it to Clephane. The promoter said "But it's not nearly as clear as the original." He handed the copy around as optimism changed to disappointment. The characters were fuzzy even though the original letters had been sharp and well-formed.

By the third try grease smudges appeared on the proofs and after two more hours with the lithographer there was little improvement. Clephane was not ready to give up. It was agreed that Ottmar should remain in Washington and try to adjust the machine.

For the next four days Ottmar, Clephane and Moore worked to obtain better results. At last Clephane had to admit that it was not the fault of the machine but of the process. The results were too irregular and unpredictable for practical use and apparently there were too many steps involved, with failure at any stage impairing the final outcome.

Clephane was unwilling to write off the machine entirely. He wanted Ottmar to build a few more like it for even if the results were not perfect, they could still be useful in certain types of work. "I think I know some people who will buy them. That way, at least, we'll get back our initial investment."

The mechanics in the Hahl shop assembled three more machines under Ottmar's supervision and these were sold to shorthand reporters in Washington, Chicago and New York, where they were used to duplicate legislative proceedings, court testimony and other documents. One office even reported that it had used the transfer typewriter to produce several pamphlets on an experimental basis, though it acknowledged that the quality was inferior. Finally, Clephane was ready to concede that the invention held out no hope of living up to his dream of bridging the gap between manuscript copy and the printed page.

CHAPTER 4

THE IMPRESSION MACHINE

In January, 1878, Clephane literally bounced into the Hahl shop. He unwrapped a package and laid the contents out. Ottmar picked up a sheet of white material that had the appearance of soft wood but was more flexible and did not have a noticeable grain.

"That is the answer to our problem," Clephane declared. It was papier-mâché, a material used by printers for making stereotype molds. The compound was tissue paper, cardboard, paste and plaster compound. Clephane went on to explain that he had been searching for a substitute for the lithography process and had stumbled on stereotyping as the answer.

The use of stereotypes in printing had been invented in 1739 by William Ged, a Scottish goldsmith. Ged found that he could make a plaster mold of metal type set up in page form. From the mold additional metal plates could be cast for printing. In this way the original type, which would normally be used for printing, would not wear out. Printers, however, were unwilling to invest in casting equipment and materials.

Later, a Frenchman named Claude Genoux invented the papier-mâché matrix, which made it possible to produce a mold simply by pressing a sheet of this inexpensive compound against the type. Then by pouring molten metal over the face of the matrix, an impression could be obtained.

Unlike handset type, the stereo plates could print a large number of copies, then be stored for future use. Identical plates could be cast from the same mold, in order to print on more than one press if speed were demanded. With the invention of high speed rotary presses which printed with rotating cylinders instead of a "flat bed," stereotyping molds could be bent for casting curved metal plates to fit on the cylindrical presses.

Clephane knew what he wanted: "I want a machine that will punch type impressions into a strip of papier-mâché instead of imprinting characters on ordinary paper. Then, instead of transferring the characters to a lithographic stone, we can make metal castings from the strip of papier-mâché."

Ottmar asked if Clephane had asked any printers about all this, but Clephane was not one to trifle over details. "Give us an impression machine and let me attend to the rest," he declared.

August said that the shop would take on the project. "But on condition that the shop not be held responsible for the final results. I cannot afford the risk of not being paid," he said. "Were you not paid for your work on the transfer typewriter according to the exact terms of the agreement?" Clephane replied. "If that is your condition we can make the same arrangements now. You will be responsible only for building a machine that is mechanically sound. Responsibility for the success of the idea itself will be mine. Even if my plan does not work—you will be paid anyhow. Provided, of course, that the machine is not at fault."

"But I must tell you," said August, "the money received for the transfer typewriter did not nearly compensate for the effort and work involved. That is what I wish to make clear." Clephane stated that if that was the only obstacle, he would better the terms. In addition to an outright cash payment for the work on the machine, August would receive a stock interest in the invention.

Ottmar decided to observe stereotypers at work. He watched as a workman brushed a thin coating of oil onto the face of type that had previously been set up and locked into a metal frame or "chase." Then the man took a thick sheet of damp papier-mâché—called a "flong"—pressed it tightly against the face of the type to receive the impression and baked it in an oven until it was hard and dry.

The papier-mâché matrix, or "mat," was laid in a "casting box" face up and a molten lead alloy was poured into it. The metal hardened at once, forming a solid plate. When it was cool enough to handle, it was removed, trimmed of burrs, and made ready to be put on a printing press. Although the whole process took no more than fifteen or twenty minutes, the skill involved was probably not lost on Ottmar. The "flong" had to be just so damp, the oven had to be heated to exactly the right temperature, the metal had to be poured quickly and evenly, and other variables had to be monitored.

In the following months Ottmar redesigned the transfer typewriter, which history would call the Rotary Matrix Machine. The biggest obstacle lay in adjusting the machine so that when the keys were depressed the type characters would strike the papier-mâché strip with exactly the right amount of pressure for an impression that was deep enough. Ottmar lengthened the levers attached to the keys, thus increasing the pressure at the point of impact between the type characters and the papier-mâché strip.

No operator could manipulate the keys with exactly the same degree of pressure each time and even a slight variation of force from key to key would result in variances in the depth of the impression made in the papier-mâché. Thus the typeface on the metal casting would be uneven in height and give poor results.

By the fall of 1878 the machine was completed. Clephane came from Washington and examined some test strips. He was, once again, overly optimistic: "This time we've succeeded. Why, with added adjustment, you should be able to obtain perfect results." Ottmar continued the demonstration and snipped the papier-mâché ribbon into small segments the width of a newspaper column and placed them in a metal frame with crosspieces for support. He placed the molds face up in a casting box and poured in molten metal. Before Ottmar was through the metal chilled and the castings were a failure. They made up a new batch of strips and placed them in the casting frame. The metal was heated until it was almost as thick as water, hoping that this would let him pour the entire casting before the lead chilled.

By the time the metal hardened, he had managed to cover the entire face of the casting frame, but when he tried to remove the slugs, he discovered that the liquefied metal not only had covered the papier-mâché but had penetrated every crevice and joint in mold. It took an hour to separate the slugs from the papier-mâché mats and more time to clean off the frame. The lead slugs had been so damaged that they were useless.

Clephane returned to Washington determined as always not to call a halt. "A few more weeks," he pleaded. "Keep at it a few more weeks." It was agreed that August and Ottmar would each receive three shares of stock, equal to a one-twenty-fourth interest in the invention. In return, the stepcousins would continue searching for a way to overcome the stereotyping difficulties.

During the summer of 1879 Ottmar worked to eliminate the problems. He came to the conclusion that the principle of the machine was basically wrong. During his three years of work on the transfer typewriter and the stereotyping machine, Ottmar had never investigated the printing field. He felt that this was part of his—and Clephane's—problem. In the fall of 1879 August Hahl, at a meeting with Clephane announced that there was no point in continuing the stereotyping experiments. The prospect for further improvement was hopeless and he could not afford to lose more time. Clephane continued to plead for one last effort to find a solution.

Clephane and the others held a meeting and decided to seek additional backing; they decided to open a small machine shop of their own in Washington and hire mechanics to proceed with the experiments where Ottmar had left off. August and Ottmar would retain their shares of stock in the enterprise. In October the stereotype machine and the casting appliances were crated and shipped to the Capital. The total payment for the three years spent in building the Moore typewriter and the stereotype machine had not covered labor costs and all August and Ottmar had to show for it were a few shares of stock and more knowledge about printing than they needed to know.

The Rotary Impression Machine was patented in 1879. Mergenthaler worked out the machine, his first invention, and received the patent in his name, but Clephane owned it. The first model used piano keys. Later the Rotary Matrix Machine had twenty-seven round keys, which operated in combination with a rotating drum. There were three wheels, each carrying complete alphabets cut in relief in steel, one for caps, one for lowercase, and one for italic. The unit was driven by a belt, and the keys operated against short pins protruding from the slots in the cylinder. A movement of a rod caused a letter in the type wheel to raise above the periphery of the rest of the wheel. The rod held the letter in position until it was pressed into a narrow strip of matrix paper, a principle similar to Moore's Double-Cylinder Typesetter. At the instant of impression the wheel ceased to revolve. The resultant strips were pasted onto heavy manila paper backing, and the matrix was ready for the casting box.

Ottmar decided to learn more about printing. He visited the Baltimore Free Library and spent time in the Schneidereith shop.

Johann Gutenberg changed the world with his movable type. His basic invention allowed the re-use of type characters, the ability to correct before reproduction, and the ability to make multiple copies. He used wine and cheese presses to bring paper and type together under pressure and to transfer ink to paper. Many referred to his invention as "artificial writing", since it replaced the laborious manual methods of the scribes. The great Bible of Mainz of 1440 or so essentially dates the beginnings of the craft of printing.

Related parts of the printing process advanced. Paper mills switched from rag pulp to wood pulp and turned out paper faster and cheaper. Between 1872 and 1892, the price of paper dropped from twelve cents to three cents per pound. Faster and faster presses came as steam power was applied with Richard Hoe's cylinder model in 1847. In 1865 William Bullock of Philadelphia constructed the first web press that could print a continuous roll of paper, increasing output from 12,000 newspapers an hour to four times that by 1890.

But the printing craft was blocked from further development by the slowness and costliness of typesetting—still accomplished by hand. This operation was the most fundamental of all, yet it had not progressed in four centuries. The printer worked from type cases which were wooden trays divided into compartments of different sizes to accommodate the different letters of the alphabet. Individual type styles and sizes, known as "fonts" were contained in two cases which were placed at a convenient height for the compositor. The lower case, contained the more frequently used small letters, was set up directly in front of the printer with its back edge raised. Capital letters were in the upper case behind the first tray. Hence the terms, lowercase and uppercase. The positions of the different letters in the case were determined by their frequency of use in the language.

The compositor placed the individual metal characters in a hand-held device called a "composing stick." It was "L" shaped with a closed side and top and an adjustable sliding end piece that could be set and tightened with a screw to establish the width of the line. The type characters were picked out one by one from the case and placed in the proper order. Words were separated by blank spaces. When the end of a line was reached, the compositor added or removed spaces between the words to "justify" the type so that it would fit the exact column width for which the composing stick was set.

After several lines of type were composed, they were moved from the stick to a shallow metal tray called a galley. Proof sheets were printed and checked against the original manuscript, and all errors corrected before the type went on the printing press. Afterward the type was redistributed in the cases so it could be used over again.

Hand composition was slow, and even though high-speed presses could be run at a rate of thousands of copies an hour, they were usually restricted to four or eight pages because of the typesetting process. Armies of hand compositors were needed to put out a large metropolitan daily newspaper. Printers and many others tried to devise various schemes to speed up typesetting. One "inventor" tied the composing stick to his waist so that he could select the type with both hands and another introduced a funnel arrangement which was supposed to catch type characters and drop them into the stick automatically.

Since 1822 and the Church machine, which was acknowledged to be the first known mechanical typesetting idea, inventors were trying to speed up the typesetting process. All failed because they tried to use handset type in some complex mechanism for either selecting, justifying or redistributing the type. The answer lay in some totally new idea. Contests were sponsored by publishers who hoped to spur the development of a practical typesetting machine. In the 1860's, the proprietors of New York City newspapers offered prizes amounting to half a million dollars for an invention that would save a mere 30% of the time and cost involved in hand composition.

In other areas of invention, vast strides were being made. Thomas Alva Edison had announced the invention of a clear bulb which converted electricity into light. The telephone now challenged the telegraph by allowing the human voice to be transmitted over wire. Westinghouse's air brake was revolutionizing railroad travel. In the 1800s the claim was made that the Patent Office should be closed down, since all the things that needed to be invented already were. That did apply to the printed word.

Clephane had tried to adapt the typewriter principle to printing. It was understandable considering Clephane's association with Sholes' invention. But typewriting and printing were different processes. Three years of experimentation had shown that combining them resulted in a curious agglomeration. If you wanted to print then you had to have raised type, period.

A machine that used printer's handset type was theoretically correct—you needed type. More than 1,500 patents had been issued and 100 models designed and, still, none were at work setting type. In fact, one of the major problems facing printers was the cost of type. Printing jobs were usually too small to warrant the use of stereotype plates and the original type was used over and over, with the need to be replaced as it wore out. This is what kept typefoundries in business.

The use of papier-mâché matrices from which the castings were made suffered because of the shortcomings of the character-by-character principle. To produce stereotype matrices, stereotypers worked with an entire page or column of type, never letter by letter or word by word. Papier-mâché mats had to be made from a mass of type, not individual characters.

A page was too big a unit to deal with, but one *line* of type could produce a reasonable matrix. The Chinese abacus must have provided an idea to Ottmar. With its parallel columns of beads sliding on wires it was a model of simplicity. A machine with a similiar series of parallel wires or bars could be built with each bar containing the type characters. The characters could move up and down, just as the columns of beads in the abacus moved. Type characters could be selected from each bar and slid down to a common level of alignment, forming a one line of a column. Then a papier-mâché strip could be pressed against the whole line to produce a matrix. There would be no indenting of individual characters one at a time, as in the earlier stereotype machine, no problem of equalizing key pressure to obtain a uniform depth of impression.

He was going to have the type characters slide freely along vertical metal bars, not unlike the beads on an abacus. But instead he simply affixed the types permanently to parallel bands which could be raised or lowered. Each parallel band would carry a full alphabet of raised type characters and various-sized space characters for use between words. Keys would raise or lower the bars which would be arrested to bring the selected characters to a certain operative level. A lever would press a damp papier-mâché strip against the line of assembled type to produce the matrix from which metal slugs could be cast.

The parallel bands, containing type characters, would be notched on the edge with teeth similar to those on a gear and a mechanism connected to the keys would arrest each band at the desired level, similiar to the watchmaker's escapement to step the mechanism of a watch or clock at desired intervals.

These plans were sidetracked when August received a notice of foreclosure. After discussions with the bank they were given another month's grace and soon after that August received word that he had won a bid to build and install an electrical clock system in one of the large Baltimore hotels. Within six months the shop was well on its way to paying off the entire bank debt.

All during the spring and summer months of 1880, the stepcousins worked hard. Business had increased and several new mechanics had to be hired. Since the firm had already established a reputation for workmanship, the combination of superior service and low prices proved to be effective in luring new business to the shop.

The "band" machine would also wait for Ottmar to marry Emma Lachenmayer, the daughter of a prominent Baltimore architect. The Lachenmayers lived in another section of town, but Emma had attended church services at the invitation of a girl friend who was a member of Ottmar's parish. They met there and he soon wrote to his parents: "I have met a young lady whom I like very much. Her family is originaly from Bavaria. I think I may marry her."

"As an engagement gift I will make you a full partner in the Hahl shop," said August. "It will begin on New Year's Day." Although there was hardly time during those hectic days to give much thought to his idea for a band machine, Ottmar regretted that he had discontinued the drawings of his "band" machine.

Ottmar and August received a surprise visit from Clephane, who had not been in Baltimore for months. They liked Clephane personally, but they had the uneasy feeling that he had a natural talent for generating more heat than light. Asked what brought him back to Baltimore, he replied: "A matter of grave urgency. You see, we are still intent on perfecting the stereotype machine. We opened a machine shop to go ahead with the work you began, Ottmar, and there has been some progress but the device is still a long way from doing what we want it to do."

He went on. "I have a proposition. I would like Ottmar to come to Washington for a day or two a month to consult with the mechanics. He is more familiar with the problems than any of us. By giving us the benefit of his experience, he can render a valuable service. Remember—you are stockholders, too. You have a material interest in the machine's success."

Clephane's mention of a fee for this service put a different light on the matter. Even August now saw it as a straight business arrangement at a fee of fifty dollars a month. Clephane accepted.

Ottmar's first trip to Washington as a consultant was also his last. The machine shop Clephane and the other backers had opened was well equipped but development of the stereotype machine was almost exactly where he had left it two years before. Ottmar spoke to Clephane and the others bluntly. "I can see no future for this machine," he reported. "The punching of impressions into papier-mâché, letter by letter, is in itself faulty. No matter how much money you will put in, the machine will never be practical. I speak to you as honestly as I can."

Clephane asked "Isn't there anything you can do? We've poured thousands of dollars into the invention already."

"I can't give you miracles," Ottmar replied. "You have paid me to travel to Washington to give advice. Well, I give it to you. I tell you to discard this machine entirely."

J.H. Crossman, one of the backers, observed, "If we were to do as you say and discard the device, we would be left with nothing, not even an idea from which to proceed." "That is true," Clephane interrupted. "You say we ought to start fresh, Ottmar. Very well. Where do we go from that point?"

Ottmar then explained his idea for a vertical band machine. "You see," he pointed out, "such a machine would have a great superiority over the stereo typewriter. It would make the impression in the papier-mâché line by line, not letter by letter. Using this principle, all of the bad points of the present machine would be overcome."

Clephane, dedicated as he was to the stereo-typewriter idea, yelled out "It's ingenious—brilliant even. Nevertheless, I think it would be unwise to discard the present machine after having sunk so much money into it without giving it every last chance for success. There is simply too much invested to give up now."

In July, Ottmar decided to sell his three shares in the stereotype machine and use the cash to help defray the expense of furnishing a new home. He wrote to Clephane, who found someone willing to purchase the stock for sixty dollars. While it was a small enough dividend for the years of effort he had invested in the machine, Ottmar decided that it was better than nothing at all.

On September 11, 1881, Ottmar Mergenthaler, age 27, married Emma Lachenmayer. After a brief honeymoon in New York City, the couple returned to Baltimore, where Ottmar settled down once more to his duties as a partner in the firm of Hahl & Mergenthaler, Machinists and Instrument Makers.

CHAPTER 5

THE BAND MACHINES

In July, 1882, while August was out of town, Clephane burst into the Hahl shop with exciting news (he always seemed to be bouncing or bursting with enthusiasm). Ottmar's band machine idea interested a man in Washington who was ready to put up real money. There was something uncanny in Clephane's coming to Baltimore at this moment when Ottmar was working on the blueprints for the machine.

"We've managed to interest a wealthy Washington lawyer in our printing project. His name is L.G. Hine. The stereotype machine was given up as a lost cause," Clephane went on, "Hine invested heavily in it before we abandoned it. But that's not the end of our plans, not by a long shot. Hine is a man of vision. I mentioned your band machine scheme. He's truly excited about it, I can tell you. He would like to see you as soon as possible."

Clephane was disappointed to learn that Ottmar was not that eager to pursue the device. He had undoubtedly expected Ottmar to leap at the offer. "But don't you see that speed is of the essence?" he urged. "We must strike while the iron is hot—while Hine is interested."

Clephane arranged the meeting for August. The man whom Jim Clephane introduced as L.G. Hine was a tall, courtly man who listened quietly and reflectively but did not say much. It was Clephane who did most of the talking. He reviewed the hopeless situation they had reached in developing the matrix typewriter and described Ottmar's idea for the band machine.

"It's a brilliant scheme, I'm convinced of it," Clephane said with his usual enthusiasim. "Mr. Mergenthaler here is a mechanical genius. We can't go wrong if we follow his plan."

Ottmar showed his blueprints and proceeded to explain the principle of the band machine. Then he went on to discuss his own personal situation, pointing out that building such a machine would be very timeconsuming and would mean dissolving his partnership with August Hahl.

"Yet you have enough faith to take that risk," the lawyer stated.

He continued "I'm prepared to back the machine personally and get friends and associates to do so, too. It will take time to get the funds together. We lost considerable capital on the matrix typewriter. But I promise you this—within a few months you will be able to proceed. The financial terms can be worked out later. Is that agreed?"

Ottmar said yes. Hine made a fleeting but significant remark "One thing more, Mr. Mergenthaler. I want you to understand that the responsibility for mechanical development will be left entirely up to you. There will be no interference from me or anyone else. You are the mechanical expert; we are the financial backers."

He soon received reports from Hine informing him that capital was already being accumulated. The lawyer suggested that he plan to commence work on the band machine shortly after the beginning of the new year. That October, Ottmar told August about the plan to dissolve their partnership in order to work on his invention. He then outlined his proposal in detail. "I would like to open a shop of my own in which to build the band machine. I will borrow money from a bank to do it. My interest in the partnership will guarantee the loan. The cost of materials and labor to produce the machine will be paid for by Clephane, Hine and the other backers."

August agreed to the arrangement. They would dissolve the partnership and he would buy out his stepcousin's interest with monthly payments which, in turn, would be used to repay the mortgage on Ottmar's new shop. In November, Ottmar took a lease on a location in Baltimore's Bank Lane section and arranged to borrow capital from a local bank to purchase machines. On New Year's Day, 1883, a sign was placed above the entrance stating:

OTT. MERGENTHALER & CO.
Mechanical Engineers and Machinists

The agreement was for Ottmar to build an experimental model first. If this machine proved successful, a full-size band machine was to follow. Hine's financial resources for the project consisted of his personal funds plus contributions from two other Washington businessmen, Frank Hume and E. Kurtz Johnson. Other backers would include T. J. Mayer, Abner Greenleaf, James H. McKenney, Samuel L. Bryan, E. V. Murphy, and Clephane.

Ottmar began work on the model immediately. The blueprints prepared earlier were so precise that the actual construction was almost a matter of routine. He hired mechanics to assist him but did most of the precision work himself. By the end of March—seven weeks after the birth of his first child Fritz—Ottmar was able to telegraph Hine in Washington that the working model of the band machine was assembled and ready for testing.

The band machine looked like a Jacquard loom to which a typewriter keyboard had been attached. Since it was an experimental model, Ottmar had kept the number of vertical bands to twelve. Mounted on each metal band was a full alphabet of raised metal type and various-sized blank characters for spacing between words. The type had been obtained from a standard typefoundry. Keys activated a series of metal stops and were ready to catch each band at the desired character. After the stops were set, a "line key" was pressed which released the bands so they could slide down freely. When the letters were caught at the proper level they were locked firmly in place. Then the operator touched a lever to press a moist strip of papier-mâché against the raised type. When the strip was peeled away it contained an indentation of the characters. The "line matrices" were dried in an oven and placed in the special casting box to produce metal slugs.

Hine came up from Washington by himself to witness the test, explaining that the other backers had been unable to take time off for the trip. He pointed out that he was empowered to act on their behalf. Ottmar demonstrated the operation of the machine, pausing to explain each step of the process and point out the various features of the mechanism. He handed metal slugs to Hine for inspection. The attorney remarked that they seemed uniformly good and announced that he would take them back to Washington to show the others. Ottmar apologized for the appearance of the machine. "This is just an experimental model," he explained. "If now I am authorized to build a full-size machine,

I will increase the number of vertical bands so that we can print forty characters on a line—wide enough for a newspaper column."

"Well, what are you waiting for, Mr. Mergenthaler?" Hine replied, "Go ahead and do it—build your machine."

There were still shortcomings. First was the problem of the stereotyping itself. While the results achieved were infinitely superior to those obtained from the matrix typewriter, Ottmar was not at all satisfied with the speed of production. The actual assembling of the characters in the machine and the indentation of the papier-mâché could be accomplished with relative speed, but it was in drying the matrices and casting the metal slugs that the process bogged down. These were time-consuming manual operations. A second problem had to do with justifying the assembled letters. The operator used a scale and pointer on the machine to estimate in advance how many blank space characters would have to be inserted between words to justify the line completely. This process took time, and if the operator calculated incorrectly, the whole line had to be reassembled.

Ottmar dealt with these problems throughout the summer months of 1883. It was not until September, when the full-size band machine was almost completed, that he finally arrived at answer—eliminate the papier-mâché matrices and substitute metal dies or "female" matrices for the raised type on the vertical bands. Line slugs could be cast in the machine itself by pouring molten metal directly onto the dies or matrices. The idea came to him on one of the trips to Washington: "Why have a separate matrix at all; why can I not stamp matrices into my type bars and cast type metal into them in the same machine?" Here we have the first combination of composing and casting in one machine.

The basic principle would remain: each vertical bar would be retained with the complete alphabet of letters, together with blanks for spaces of different widths. The difference would be indenting the characters on the bands, instead of having them raised, thus serving as direct matrices, which would eliminate the troublesome papier-mâché matrix and hand casting of slugs. There would be a heated pot for the molten lead alloy and when a line of matrices was assembled and justified, a mold could be brought against it and molten type metal forced through from the liquified metal pot.

This was the principle used by typefounders to cast metal

type. In the typefoundry, each female matrix served as a mold for an individual letter. When type metal was poured into the mold, it solidified almost immediately. In this machine, the casting would be done a line at a time, instead of letter by letter.

The solution to the "justification" problem was as simple as bottle stoppers, which are designed with a slight taper so they can be pressed down to give a tight fit. The operator of the band machine assembled a line of characters from the keyboard, and instead of calculating the size of the space between words, he brought the line to column width by means of wedge-shaped word spacers. The wedges were pressed down to expand the line to the left and right margins and make it fit like a cork in a bottle. In effect, it would provide a system of automatic justification. The time spent by the operator in mental calculation would be eliminated. This was the first version of the first band machine.

Neither of these improvements could be incorporated into the present machine because the final assembly was too close to completion. Ottmar decided to complete the present model, show it to Hine and the other backers and convince them to sponsor a new version with the direct casting and automatic justification features.

In October, Hine, Johnson, Hume and Clephane came to see the band machine in operation. The mechanism worked perfectly. Everyone examined the papier-mâché matrices and finished metal line slugs and expressed enthusiasm at the results. Clephane, as expected, was the most extravagant in praise: "This time you've done it. The machine will revolutionize printing, no question of it."

"It is a development in the right direction," Ottmar said. "But we must go a long way yet before we reach our final goal." He went on to explain his plans for the direct-casting machine and the wedge justifier. "Now such a mechanism that will do away wth papier-mâché and justify each line without effort on the part of the operator—that will truly be a practical step. I see our past work and pains as a further advance to the final destination. "But to stop here and be satisfied with a half-success might lose us everything."

Hume and Johnson were reluctant to approve Ottmar's request for funds to build a new machine. "Why should we invest more money without first exploiting the commercial possibility of this model?" Hume wanted to know. At this point Hine stepped in to make the decision. "Gentlemen, your desire to realize dividends from what we have invested thus far is understandable from a businessman's point of view," he told his fellow backers. "But Mr. Mergenthaler is right. We must take the long view. A desire for a quick profit now might compromise larger gains in the future. Of course we will go along with our young colleague's proposal. I advised him at the start that he would be the final arbiter in technical matters. Until now his judgment has been sound. I see no reason to alter our initial agreement at this point."

Hine now turned to Ottmar and said, "We are severly limited in funds. For that reason, Mr. Mergenthaler, I suggest that the improvements in the new machine be limited to the direct-casting feature. If the model is successful, I personally will guarantee the backing for a second machine with automatic justifier."

Hine proceeded to outline a generous agreement to cover Ottmar's interests. While all inventions or improvements were to become the property of the National Typographic Company, the inventor was to receive a 10% royalty on every machine manufactured by or for the company. "In addition, Hine added, "as soon as a machine suitable for practical use is produced you will receive a thousand shares of capital stock. You are also to be employed as salaried manager of all shops and factories, with final authority and responsibility for their operations."

The National Typographic Company was incorporated in December, 1883, and registered in West Virginia. The machine on which it was based was far more complicated than the first band machine since the introduction of a casting unit multiplied the number of mechanical operations. While the original model had been worked entirely by hand, Ottmar decided to power the new machine with steam. The vertical metal bands which formerly held raised type characters now would contain indented, or female, characters, or matrices. After the operator pressed the keys the bands would descend to different levels, so that the desired matrices would be brought into line, just as in the earlier model. At this point, however, the casting mechanism would take over. A sliding mold would cover the line of assembled character matrices. Molten metal would be forced into the mold from a pot constantly heated by a gas flame, through a pump arrangement. Then the mold would slide back, the metal casting or slug would be ejected through knives which would trim it of excess metal, and the finished casting would drop into a proof galley. The vertical bars would be lifted to their original position, ready for the next line.

The incorporation of the casting system was no problem. The problem involved the cost of the female matrices. To obtain the type for the earlier indenting machine he had simply gone to the typefoundry and purchased raised type characters, handset type, just as any printer would have done. He had assumed that the cost of the female matrices, from which raised type was cast, would be equally low. He was dismayed to learn that each matrix could cost $2 each. The craft of typefounding also had changed little since Gutenberg's day.

It was an art that depended largely on the hand skills of highly trained workers. Each typeface was the creation of an individual designer who prepared large drawings of the letters of the alphabet and other type characters. From these basic designs, experienced craftsmen carefully engraved sets of steel punches, with each punch containing a raised character image in a particular size. The punches, in turn, were employed to hand-produce matrices of steel or brass from which printer's type could be cast in quantity, through the use of a lead alloy.

Since each band of the line casting machine would have to contain the twenty-six letters of the alphabet plus other type characters, Ottmar estimated that forty-five hundred individual matrices would be needed for each machine. The total would be nine thousand dollars, not including the cost of manufacturing the line caster itself. He knew that under such circumstances the entire project was out of the question commercially. He and Hine had agreed that any machine would have to sell for no more than $400 in order to win practical acceptance by the printing industry.

He considered stamping the characters into the bands with type punches. Steel type punches, from which matrices were made, cost about five dollars each and could be used over and over for stamping female characters into the metal bars. And since the major cost of producing a matrix was in finishing the metal block rather than in the stamping, he could punch matrices into the bands of the machine at low cost.

In spite of apprehension, the plan did work. Ottmar purchased sets of matrix punches and set a mechanic to work stamping the characters into the matrix bands. Throughout the winter and spring months of 1884 he and his men worked to finish the machine.

In February Emma gave birth to a second son, Julius. Ottmar observed that the first indenting band machine had been completed in honor of Fritz. The line casting band machine would probably be for Julius.

By June Clephane was writing almost daily, inquiring when the model would be ready for a test. That day finally arrived and all the Washington backers showed up as well as local spectators, including the Schneidereiths, who had heard of Ottmar's "magic machine" and were curious to see it in operation. Even August Hahl was there.

Ottmar, who had been up all night making final adjustments, was still hard at work when the crowd began to collect in his shop. "Everything is ready, ladies and gentlemen," he announced.

The crowd watched as Ottmar sat down at the keyboard and tapped out a line. Then he slowly rotated the heavy driving pulley by hand to check the action of the mechanism. The matrix bands slid into position. The machine clanked flawlessly through the entire operating cycle.

"It seems that the mechanism will not explode after all," he observed. "Now that we have completed the rehearsal we can go through the actual performance."

He attached steam power. Then he reached over and removed the stopper from the metal pump which had been heating over a gas flame. He tapped out a series of letters on the keys, touched the "line key" and leaned back with his arms folded. The vertical matrix bands slid into place. They were clamped and aligned. Then the metal pump discharged its contents into the sliding mold and in a few seconds a finished line slug, gleaming like silver, dropped from the machine's innards. In the meantime the matrix bars were hoisted to their normal position, ready for the next line to be composed.

The entire operation had taken only fifteen seconds from the time he had tapped the line key. From that point on, he had not touched the machine again. Every operation had been carried out automatically. The crowd applauded. Clephane, still naturally high, repeated over and over, "Amazing. Positively amazing."

The line slug was passed from hand to hand and examined by each of the spectators. Ottmar now asked Julia Camp, a typist whom he occasionally employed for his business correspondence, to take over. He had shown her how to operate the line casting machine the day before. She sat down at the keyboard and for the next hour produced line slugs containing brief messages which some of the spectators had written out and asked to have composed. She also copied two paragraphs from a newspaper article, handed the slugs to Ottmar who inked them and rolled a proof. The type impression was perfect.

Hine looked pleased as punch as he examined the proof sheet and showed it to the other backers. They nodded and held a football-huddle conference. "We are all agreed that your machine exceeds our highest expectations," Hine announced. "We want you to proceed as soon as possible to build a second model containing the automatic justifier."

CHAPTER 6

THE INDEPENDENT MATRIX MACHINE

While Ottmar took on the task of perfecting the wedge justifier, Hine and Clephane were laying the groundwork for commercial exploitation. As soon as they returned to Washington they launched a campaign to publicize the Mergenthaler linecaster (it now became one word) and secure additional capital. It was one thing to underwrite the construction of an experimental model, and quite another to set up a factory to produce reliable machines in quantity.

Although their personalities contrasted sharply, Hine and Clephane were an effective team. The lawyer lent an air of quiet dignity as the official spokesman for the National Typographic Company while Clephane, ever enthusiastic, showed himself to be a persuasive salesman. By December, 1884, the number of stockholders had increased to almost a dozen and a stock issue was floated. Among the backers was Stilson Hutchins, publisher of the *Washington Post*.

Ottmar was summoned to a meeting of the directors in Washington. He reported that progress on the automatic justifier was proceeding rapidly. "There is no reason why it should not be ready by February," he said. On returning to Baltimore Ottmar went to work to locate a suitable factory site. With August Hahl's help he found an empty building at 201 Camden Street.

Ottmar arranged to have the equipment moved from Bank Lane to Camden Street at night, so that there would be no time lost. He also placed orders for a selection of additional machine tools and equipment and explained to Hine that obtaining the most up-to-date machinery now would save time and money later on when they were ready to go into commercial production. Hine agreed with the soundness of this philosophy and gave him authority to proceed as he saw fit.

Toward the end of January he telegraphed Hine in Washington that the machine was ready. The lawyer's reply was cryptic: "Ship machine to Chamberlain Hotel, Washington, at once. Come yourself next Monday. Be prepared to stay for a week." Ottmar wanted his wife to accompany him, but Emma was expecting their third child.

When Ottmar arrived at the Chamberlain Hotel he discovered that Hine and Clephane had taken over a large suite and set it up as a special exhibit area for the Mergenthaler linecaster. For the next five days, there was a constant stream of visitors eager to see the machine in operation. The hundreds of spectators included printers, potential inventors and reporters assigned to cover the event for their newspapers. In addition, a number of important government officials were on hand.

One afternoon a messenger rushed up from the lobby of the Chamberlain Hotel with instructions from Clephane that the exhibit suite be cleared at once. A special party was on its way up to see a demonstration of the machine, Ottmar was told. A few minutes later a distinguished-looking group entered the suite, accompanied by Clephane, Hine and several other backers. "I have the pleasure of introducing the President of the United States," Clephane announced.

President Chester A. Arthur was a big man with silver gray hair and a flowing mustache. He was about to see a strange-looking machine. He approached the linecaster and examined it with interest, asking a number of questions about its operation. "I would like very much to show you how the machine works, Mr. President," Ottmar said.

He sat down at the keyboard, tapped out a line and pressed the line key. President Arthur stared as the metal bands clattered into place and gears began to turn. A few seconds later a shiny metal slug was ejected from the machine and fell into place in the galley tray. The President examined the slug and said "This is truly a marvelous device. You are to be congratulated on your splendid achievement, Mr. Mergenthaler. I hope that you will reap proper financial reward for your labors."

There was a banquet and the last speaker on the program was Ottmar himself.

Allow me, gentlemen, to express my hearty thanks to you for the honor you have bestowed upon me in coming here to witness the performance of my invention. You have come here to witness the operation of a new composing machine, and in as far as we are working in a field which is strewn with the wrecks and failures of former efforts in the same direction you will probably ask, "Are you going to have more success than those who have gone over that field before you; and if so, why?" My answer is, "Yes, we are going to have full success for the reason that we have attacked the problem in an entirely different way than did those who have failed."

When I started on this problem I surveyed the field and selected the best road, regardless of the roads which others have taken. I knew the direction in which others had attempted to solve the problem, and was careful not to fall into the same rut which had led every previous effort into failure and ruin. We make and justify the type as we go along, and are thereby relieved from handling the millions of little tiny types, which have proved so troublesome to my predecessors who have failed. We have no distribution, yet we have a new type for every issue of a paper, an advantage which can hardly be overrated.

I am convinced, gentlemen, that unless some method of printing can be designed which requires no type at all, the method embodied in our invention will be the one used in the future; not alone because it is cheaper, but mainly because it is destined to secure superior quality.

The history of our enterprise, gentlemen, is one of evolution. We started by printing one letter at a time and justifying the sentences afterwards; then we impressed into papier-mâché one word at a time, justified it, and made a type from it by after process. Next we impressed a whole line and justified it, still leaving the production of the type as a second operation; but now we compose a line, justify and cast it all in one machine and by one operator.

> It is a great result, but, gentlemen of the Board, to you it is due as much as to me. You have furnished the money, I only the ideas; and in thus enabling me to carry this invention to a successful end you have honored yourselves and your country.
>
> I say you have honored your country, for everyone will know that this invention has been originated in the land which gave birth to the telegraph, the telephone, the Hoe press and the reaper; everybody will know that it came from the United States, though comparatively few will know the name of the inventor. Gentlemen, again I thank you.

Hine and the other stockholders agreed on the wisdom of putting the linecaster on the market as soon as possible, while it was still fresh in the public mind. By mid-March production plans were nearly completed. His work was interrupted when Emma gave birth to their son, Eugene.

There were two fundamental defects in the present model, he decided—both having to do with the speed of operation. Under the present system, it took fifteen seconds from the moment the line key was pressed until the finished slug of type characters dropped out of the machine. During that time the operator had to remain idle while the machine did its work, a waste of time and labor. How much more efficient it would be if the operator could assemble lines of matrices continuously, as fast as he could press the keys, with the actual casting independent of the operation of the keyboard. The rate of production of the line slugs could be increased three of our times.

His second concern grew out of the difficulty of changing typesize or style in the linecaster. In order to replace a font the operator had to laboriously remove each of the forty vertical metal bands and substitute a different set of bands containing the desired matrices. This was a tedious, time-consuming process.

It was the vertical bands themselves that constituted the difficulty. As long as the operator had to wait until the line was cast so that the bands could return to their original position before assembling the next line, a time lag was inevitable. There was no conceivable way of changing type style or size without changing the bands—unless each machine were to contain many sets of bands. But that would make it too cumbersome and expensive, and useless from a practical standpoint.

The solution that finally came to him was the same idea that had occurred to him when he had first had the notion of building a vertical band machine. Why not use single character matrices that could move independently instead of vertical metal bands with fixed columns of matrix impressions? He had reverted to the old scheme of the sliding beads on the abacus.

He recalled that his first notion had been to attach individual type characters to parallel wires so that they could slide up and down freely, adopting the abacus principle literally. Instead of brass type bars, why not have a series of hollow vertical tubes or channels, each with a supply of independent female matrices for a particular type character.

A spring mechanism at each channel would be activated by a finger key and would release an individual matrix. A line would be composed of these independent matrices. While the machine was making a casting of one line, the operator would be able to assemble the next line. Both processes would take place simultaneously. Thus, the production of the line slugs would be continuous and would proceed as fast as the operator could depress the keys.

With independent matrices, it would be possible to store them easily in removable magazines. The storage containers or magazines would be little more than large flat metal cabinets divided into narrow vertical channels. Each magazine would store a font of type. When a change of type style or size was desired, one magazine would be slipped out of the linecaster and another substituted.

But how would the individual matrices be redistributed after the line slug was cast? They would have to be returned to the matrix magazine so they could be used over again. A recollection of a somewhat similar problem some years before concerned a sorting machine. Items, such as fruit, passed over a series of openings of various sizes. The holes increased in size along the way and as each piece of fruit came to the proper opening, it dropped. Type matrices were not fruit and could not be sorted by size. Why not choose some other characteristic for sorting? Locksmiths were able to shape keys to fit individual locks so that the notches and grooves would "code" a key so that it would fit one particular lock and no other.

Ottmar introduced notches and grooves into the molds so that they looked like keys. The face of each matrix would contain an indented female type character. Once the matrix line was assembled by keyboard operation and the casting completed, the matrices would be elevated automatically by a pulley arrangement to the top of the machine and passed over a series of openings, one for each channel in the magazine. Every opening would be notched and grooved differently. When a matrix with related notches and grooves passed over the correct opening it would fall into its proper channel ready to be used again.

Each type character would have a standardized "coding," regardless of type style or size, so that magazines could be interchanged at will. In spite of the complex problems presented by the independent matrix principle, Ottmar managed to work out the basic theory in less than three hours time.

There was only one serious obstacle—the cost of the matrices themselves. When he had first sought to buy matrices to attach to the vertical bands on the linecaster, he had been dismayed to find that they cost two dollars apiece. Since thousands of matrices were needed for each machine, the cost had been prohibitive. Therefore, out of necessity he had devised the scheme of buying type punches and impressing the type characters in the vertical bars.

He calculated that under the new design he would need twelve hundred female matrices for each linecaster. Twelve hundred would cost much less than forty-five hundred but it was still too costly. Since each matrix would have to be grooved and notched in a special way, there would be added expense.

Ottmar cut notches in a sample matrix and took it to John Ryan, one of Baltimore's largest typefounders. He told Ryan about his idea for the new machine and indicated that he would need to order matrices in considerable quantities. "They must cost no more than six cents apiece—for that is what I have calculated to be the greatest cost at which I can produce the machine economically," he explained.

Ryan handed Ottmar a standard matrix produced in his own foundry and said, "This matrix is a much simpler piece of work than yours, and it costs a dollar apiece to adjust—after the type impression has been made. If you can produce these matrices for typefounders at fifty cents apiece, there is a fortune in store for you, I assure you. And that is without regard to the success or failure of the linecaster itself."

Ottmar made the rounds of other Baltimore typefounders. In each case the answer was the same. He knew that the main item of expense in producing matrices at the typefoundry was the cost of labor. As long as matrices were made by hand, Ryan was probably correct in stating that the cost of production could not be reduced drastically. But couldn't a steam-driven stamping machine be used to produce them in quantity? Two of these machines had just been delivered to the shop and were waiting to be uncrated.

For the next few days, Ottmar spent all of his time cutting a matrix die out of a standard steel type punch. Staring through his loop, he finished the die by hand, using the tiny, high-precision watchmaking tools which he had mastered as a youth. The die was now locked into one of the stamping machines and several matrices were punched out of a small sheet of brass. Ottmar measured the dimensions of one of the samples with a micrometer. Every matrix measurement was accurate to a hundreth of a millimeter.

He arranged for an immediate conference in Washington with the directors of the National Typographic Company. Using a large drawing he explained his new design in detail. "What I ask of you gentlemen is a decision to postpone commercial production of the vertical band linecaster until a model of my new design can be completed," he said. The directors were disappointed. They had looked forward to the start of quantity production. Now, with the promise of commercial success clearly within their grasp, they were being asked to wait some more.

Hine, the chairman, said, "Not many stockholders can stand being told that we have the best machine in the world, but that we are going to make another which is still better."

Ottmar replied that he was no more certain that the independent matrix machine would work than his own belief in its feasibility. He estimated that it could be built in about five months. As for the matrices, he felt sure that he could mass produce them at a small fraction of the commercial cost. He even passed some of his sample matrices around and announced "These were made in our own shop with our own machines and our own dies."

Stilson Hutchins looked at Ottmar's drawing and mused, "I am no engineer, Mr. Mergenthaler, and I have the utmost confidence in you. But as I examine this sketch and listen to your description of what the machine will do, I can only shake my head in wonder. Why, you are asking us to believe that you can construct a mechanism with a brain of its own."

Ottmar answered, "I suppose one might call it a machine with a brain—a brain of brass and steel. But I speak with the greater sincerity, gentlemen, when I tell you that if I had not first convinced myself of the practical possibility of building such a machine, I would not have brought the proposal before you today."

Ottmar, who was not a director of the company, was now asked to leave the room so that the board could meet in executive session. For the next hour and a half he sat outside while debate went on behind closed doors until he was recalled.

"Mr. Mergenthaler, I know you will be pleased to learn that we've endorsed your request," Hine told him solemnly. "You have six months to show results. We owe you that much. If, at the end of that time we are convinced that no progress is being made on the independent matrix machine, we'll proceed to manufacture the vertical band model as planned."

CHAPTER 7

THE PUBLISHERS SYNDICATE

The independent matrix machine was similar to the band model in its casting principle, with a fundamental difference in the replacement of the bands by a magazine filled with individual matrices and a mechanism to redistribute the matrices. The original plan called for each brass mold to be released from its channel by keyboard operation, so that it could drop into an inclined chute and slide down to its proper place in the assembling line of matrices. Individual matrices would stick in the chute and refuse to slide freely, thereby disrupting the whole assembling process. He solved this problem through the use of compressed air. As a matrix was released from the magazine by a touch on the proper key, it was carried by an air blast along the inclined chute to the assembly point.

The new design began to take shape, but construction of the machine itself was only one of the tasks before him. Just as important was the creation of a system to produce the tiny brass molds in quantity. The matrix was the key to the success of the venture. Without a source of supply, the machine was useless.

He came to the conclusion that a separate factory would be required just for the manufacture of the matrices. Some thirty machines would be used for the various stamping and finishing processes, and although the initial investment would be high, but it would guarantee a continuous flow of molds at less than the six cents which he had set earlier at the maximum cost per matrix. That goal was met. In 1915 the average price of a matrix was fifteen cents and in the 1960s it averaged thirty five cents to the buyer, but cost less than ten cents to make.

Through the efforts of Stilson Hutchins, some of the nation's leading newspaper publishers were becoming aware of the revolutionary events in Baltimore. Accompanied by Clephane and Hutchins, a number of publishers came to Camden Street to see the band machine and to hear about the new circulating matrix design. They included Whitelaw Reid, publisher of the famous *New York Tribune*.

Reid was quite famous. He had been a correspondent during the Civil War and covered the major battles. His report on the death of Lincoln was the one most people were familiar with. His work led Horace Greeley to make him managing editor in 1868, and after Greeley's death, Reid gained control of the newspaper.

"This is the miracle of the age," he told Ottmar. "You'll harvest fame and glory, I promise you. Some day in the not too distant future you will return to your fatherland a rich man. You and your family will occupy one of the most noted and beautiful castles on the Rhine."

Ottmar thanked him but observed that he was not interested in returning to Germany. "Though I was born there, I now consider myself an American," he said.

In May, an eventful meeting took place between a small group of publishers and the officers of the National Typographic Corporation. The former offered to pay $300,000 for a controlling interest in the linecaster. It was an enormous sum, enough to enable the company to launch production on a commercial scale. Hine and his colleagues accepted the offer and plans were drawn up at once to organize a new company, to be known as the Mergenthaler Printing Corporation. It was to be a "parent corporation," combining the stock interests of the National Typographic Corporation and the new syndicate of publishers. The members of the syndicate were Whitelaw Reid, W.N. Haldeman of the *Louisville Courier-Journal*, Victor Lawson and Melville Stone of the *Chicago News*, Henry Smith of the *Chicago Inter-Ocean*, W.H. Rand of Rand, McNally & Company, Chicago, and Stilson Hutchins of the *Washington Post*.

With the passing of control from the Washington group to the publishers, an election of officers was held. Melville Stone was named president to succeed Hine. Though the new company had been named in his honor, Ottmar did not participate in the complicated business transactions. He was not an officer of the company and had no direct voice in the decisions of its directors.

He was busy trying to meet the deadline for the new machine to attend frequent business meetings in Washington.

In mid-June Ottmar was called to a director's meeting in Washington to deliver a progress report. There were many new faces seated around the conference table, and the atmosphere was more formal and businesslike than in the past. He made his report brief and to the point. Construction of the independent matrix machine was proceeding rapidly, he said, and unless an unexpected difficulty arose, it would be ready for a test in late July—a month earlier than the deadline.

Melville Stone thanked him on behalf of the directors and went on to the next order of business. "As chairman," he said in a matter-of-fact voice, "I now entertain a motion that the Mergenthaler factory be moved from Baltimore to Chicago. Such a move is essential if I am to carry out my duties as president."

Ottmar protested. "I know nothing of this plan," he said. "No one told me about it before. I do not see what the nearness of the shop to the president has to do with bringing out a practical linecaster."

Stone was taken aback by Ottmar's outburst, stating that as the company head it was necessary for him to supervise the shop's activities.

Ottmar replied that this argument was nonsense. "Are you a mechanic or an engineer, Mr. Stone?" he demanded. "For that is what you would have to be to 'supervise'—as you call it—the operations of the factory. If you are not technically qualified, then I cannot see any purpose in such a move, particularly at this moment when we are working against time."

For the next hour the debate went back and forth. Ottmar declared flatly that he had no intention of uprooting his family and himself from their home and friends. "If that is the price I must pay to continue my work for this company than I am prepared to resign immediately," he threatened.

Someone suggested that a vote be taken and the motion to move the shop to Chicago was quickly defeated. Stone had been voted down by his board on a key issue. It was hardly an auspicious way to begin his tenure as president of the Mergenthaler Printing Company.

By mid-July the independent matrix machine was completed—a month ahead of schedule. Ottmar invited the officers of the company to Baltimore for a private demonstration. The machine worked perfectly. When Ottmar tapped the keys the small brass molds were released from their channels, and a blast of escaping air blew each matrix down the inclined chute until it appeared in the assembly position, in full view of the operator. A tap on the space key dropped a wedge space band between words to justify the line. Finally, by pressing the line bar on the keyboard Ottmar caused the entire line of matrices to travel automaticaly to the face of a slotted mold where metal was forced into the depressed characters. While this process was taking place, Ottmar was keying to assemble a new line of matrices.

Once the silvery metal casting was trimmed and ejected, the matrices were transferred to an elevator and raised to the top of the machine. Here they passed over the distributor apparatus which returned them to the vertical channel in the magazine, ready for use again. It could almost think for itself, this complex maze of cams, belts and gears. The process was so completely automatic that the operator had nothing to do but devote his full attention to the keyboard. Ottmar demonstrated how swiftly the machine could work by tapping the keys as rapidly as he could. The metal slugs began to drop into the galley tray below the ejector mechanism with a rhythmic clank. "See," Ottmar declared, "it makes castings as fast as I can work the keys."

The officers were so impressed with the demonstration that they were in favor of going into commercial production immediately. But Ottmar once again tried to be more realistic. "The model was built hurriedly," he protested. "There are mechanical improvements that I want to make before large amounts of money are risked to produce it in quantity."

Melville Stone insisted that there was little to be gained by delay. Ottmar pointed out that he has entitled at least to the full six months which the previous board of directors had granted him. "Why should I be penalized for completing the machine earlier than the deadline?" he argued. Stone was then interrupted by Whitelaw Reid. "Very well, Mr. Mergenthaler," the publisher of the *Tribune* said, "I for one am prepared to accede to your request for a delay. I propose that you be given until October to complete your improvement."

Reid was one of the heaviest investors as well as the publisher of one of the nation's leading newspapers, and carried a great deal of weight, and Melville Stone did not oppose the motion. A vote was taken and passed unanimously.

The publishers who now controlled the company were a breed apart from Hine, Clephane and the other early backers. The latter were investors, but also visionaries, ready and willing to wager on a dream—their dream and Ottmar's dream. The newspaper publishers, on the other hand, were businessmen motivated primarily by the desire for a profit.

In early fall, a reorganization of the Mergenthaler Printing Company took place. Melville Stone stepped down and was succeeded by Whitelaw Reid, who was to serve as president and general manager of the corporation. At the same time, new capital was coming into the company at an amazing rate—largely through the influence of Reid. The October meeting of the board of directors opened on an optimistic note. Reid, who now served as chairman of the meetings, announced that the assets of the company had reached a million dollars. He then turned to Ottmar and asked if he had completed his improvements. "I have made those that I feel were most important," Ottmar replied.

"Then you feel we can begin commercial production?" Reid asked.

"Well, there are some things I would still change. . ." Ottmar began. Melville Stone jumped to his feet, "Come, come, Mergenthaler, let us have a direct answer for once," he challenged. "We are businessmen, not philanthropists. You have demonstrated that the machine works, and we have given you the additional time you requested. Now let us get on with it. I dislike this cat and mouse game."

Ottmar replied "I do not mean to sound evasive, Mr. Stone. It is just that I have the desire, as I am sure you have, to put out the best machine that is possible to make. However, to answer the question directly—yes, I think that we can now think of manufacturing the linecaster, provided we limit the number to a modest quantity."

"Good," Stone said. "Then I propose we begin with one hundred machines." Ottmar was shocked. "I said a modest number. One hundred is hardly that."

"You mean we haven't the facilities to turn out a hundred machines?" Stone demanded.

"We can build them," Ottmar explained. "But I think it

would be a serious mistake to begin with such a large number. The machine must still pass the hardest test of all—practical operation in the printing shops. Undoubtedly there are many weaknesses that will only show up after continual use. That is why we would be wise to make a small number and have them tested over a length of time—say six months or a year. Then we can make final improvements before building the linecaster on a larger scale."

Stone literally exploded. "I am tired of listening to this counsel of delay and more delay. I have considerable funds in this enterprise, and I don't propose to wait a year or more before there is even a chance for a return on my investment."

In the debate that followed, the directors were split into two distinct groups. Stone and the other publishers demanded that production of a hundred machines began at once. Hine, Clephane and the rest of the Washington group backed Ottmar. "Those of us who have worked with Mr. Mergenthaler for a number of years have the utmost faith in his judgment," Hine said. "He is the technical expert. Gentlemen, if he urges a conservative approach, I say we should take his advice."

Whitelaw Reid took no active part in the debate. He sat back and listened to the arguments on both sides. Finally, he turned to Ottmar and inquired, "When you say you are in favor of producing a modest number of machines, precisely what figure do you have in mind?"

Ottmar had discussed this with others and thought six would be an ideal number for testing purposes, but to mention such a small figure would further infuriate Stone. So he decided to compromise. "I would say a dozen, no more," he replied.

Whitelaw Reid declared: "I will support you, Mr. Mergenthaler. Shall we take a vote?"

Once again Stone was to have Reid—a fellow publisher—desert him in favor of Ottmar. The New Yorker's decision again settled the issue and the vote was a formality.

Afterward Reid asked Ottmar, "Do you understand that you are now to proceed with construction of twelve linecasters at once and to complete them as soon as possible?" Ottmar agreed.

"Very well, the meeting is adjourned."

The task of producing a dozen linecasters was not simple. It meant ordering more equipment, hiring more men and organizing a system that would keep the cost of each machine as low as possible. Moreover, even with production under way, Ottmar—always the perfectionist—insisted on incorporating mechanical improvements as they occurred to him.

Simultaneous with the assembly of the machines themselves, he plunged ahead with quantity manufacture of matrices, for without the small brass molds the linecaster was like a gun without shells. The plans he had drawn up a few months earlier for a matrix factory were put into operation. With space in the Camden Street shop at a premium he rented another building on Preston Street and hastily installed thirty special machines for the various stages in the production of precision matrices from rolls of brass. There were machines for punching blanks, machines for scraping, milling, grinding, cutting and polishing, presses for stamping the matrix characters, machines for cutting the matrix teeth and for cutting ridges on the distributor bars of the linecasters, and many other machines to perform highly specialized tasks.

By late June of 1886 the first of the dozen linecasters was ready. The directors of the Mergenthaler Printing Company agreed that it would be fitting to have the first machine installed at the *New York Tribune*, the newspaper published by their president, not unmindful that initial use by one of the nation's leading papers would be valuable in publicizing the machine.

On July 3, 1886 the composing room of the *Tribune* in New York City was crowded with visitors. Newspaper officials, reporters and representatives from leading printing firms were present. It is said that even the *Tribune* printers had dropped their own work to witness the operation of the strange machine which was supposedly going to replace them. New York newspapers were not strangers to technology. On August 31, 1861 Horace Greeley converted the *Tribune* to printing from curved stereotype plates. In 1880 the first halftone photograph, based on the photoengraving process developed by Frederick Ives of Philadelphia, was printed by the *New York Graphic*. An in 1882, the *New York Herald* installed the first Hoe "double supplement" press.

Whitelaw Reid, self-assured and elegantly dressed, stood with Ottmar. Reid pointed to the linecaster and said, "This gentlemen, is the machine about which you heard so much." Then he placed a hand on Ottmar's shoulder and added, "And this is the young man of thirty two who produced the miracle—

Mr. Ottmar Mergenthaler." There was polite applause.

Ottmar removed his coat jacket and sat down at the keyboard. He turned on the steam power and pecked at the keys for several seconds. In a few moments a metal slug bearing several words of new type appeared at his elbow. Reid picked up the first slug of metal, examined it and supposedly exclaimed in a voice calculated to have as dramatic an impact as possible, "Ottmar, you've done it. A line of type."

Ottmar continued to produce castings as rapidly as he could key. The first operator after Ottmar was John T. Miller, a member of the New York Typographical Union, who had come to the paper as a hand compositor. The slugs were passed out for inspection as soon as they dropped from the machine. Reid and Ottmar were asked dozen of questions about the linecaster while the newspapermen present took notes. A reporter asked, "Mr. Reid, have you selected a name yet for Mr. Mergenthaler's 'line-of-type' machine?"

"The answer is obvious," he replied. "We are going to call it the Linotype."

CHAPTER 8

THE MERGENTHALER LINE-OF-TYPE MACHINE

In late July, Ottmar received an official directive from Whitelaw Reid ordering him to increase the original order by a hundred. Instead of making twelve machines, he was to produce one hundred and twelve.

The policy agreed upon by the directors called for the manufacture of an experimental lot of a dozen Linotypes. Reid himself had taken the initiative in supporting such a plan. Ottmar rushed off a telegraph message to New York asking for an explanation. He received a brief formal reply from Reid stating that the earlier instructions were correct. No reasons for the change were given.

Ottmar immediately told Hine and Clephane of the situation. They demanded that Reid call a formal meeting of the directors. It was a different Whitelaw Reid who presided at the meeting. He was now cold and aloof. Ottmar explained that to manufacture a hundred additional machines would cost at least $130,000. Staking such a sum on the result of a single Linotype which had been in practical operation only a few weeks was foolhardy. But Reid and the other publishers ignored this advice. "At least reduce the number to twenty four," Ottmar asked and this suggestion was rebuffed.

The issue was put to a formal vote. With the publishers holding a controlling majority, there was little doubt of the outcome, even though Hine, Clephane and the other Washington directors supported Ottmar. Reid was upheld.

Ottmar returned to Baltimore to prepare for an expanded production program. Whitelaw Reid had concurred with Melville Stone and the other members of the publisher's syndicate, knowing full well that commercial production could never get started without Ottmar, who must have thought about resigning and starting from scratch but then realized that all the patents were held in the name of the Mergenthaler Printing Company.

By the end of 1886, the first lot of a dozen Linotypes were completed and delivered to the *New York Tribune*, and work was begun on the additional hundred machines. The introduction of the Linotype at the New York newspaper was already making an impact. Most of the columns of the *Tribune* were now being set by machine and a book called "The Tribune Book of Open Air Sports", 500 pages in all, composed entirely by Linotype, was published.

With outside orders starting to come in, Whitelaw Reid demanded to know when additional Linotypes would be ready for delivery. Ottmar had other things on his mind which he considered of greater importance because reports from the composing room foreman and Linotype operators at the *Tribune* on the performance of the first twelve Linotypes indicated that a number of mechanical changes could be made to improve the machine. It was the information he had been hoping to receive, and he was grateful for every scrap of data that was sent to him. Even while the new machines were being assembled, he began to incorporate a number of alterations in the design.

The messages from Whitelaw Reid grew more insistent. In March of 1887 he called a special meeting of the directors. Reid proceeded unceremoniously to set down dicta for Ottmar to obey. "You are to cease and desist at once from carrying into effect any changes or improvements in the design," the publisher ordered. "The machines will be manufactured exactly as they are presently. They require no further innovations. I know, for the Linotypes in my composing room are entirely satisfactory. Moreover the order for one hundred machines is to be increased to two hundred to fill the outside orders that are mounting up, is that understood?"

Ottmar begged the directors to allow him to incorporate at least a few of the more important changes. "The Linotype is not just a machine to me. It is like an organism that must be permitted to grow and improve as it goes along; it is like flesh and blood to me."

If he was incapable of turning out the machines more rapidly the company would find someone who could, Reid warned.

The following week, Emma gave birth to a fourth son—Herman. He was born the same year as the Linotype and the only child to live to see photographic typesetting and the technologies of the nineteen sixties. His son, the last male heir to carry the Mergenthaler name, would die during World War II.

To speed up production of the present model—named the "Blower Linotype" because of the use of compressed air to move the matrices—Ottmar established contracts with several independent firms to construct frames and other large parts. He increased the work force from 40 to 160 employees in the Camden Street factory and to over a hundred in the matrix shop on Preston Street. Reid continued to send messages urging faster delivery.

By February, 1888, more than fifty of the first order of one hundred had been delivered to the *New York Tribune*, *Louisville Courier-Journal*, *Washington Post* and *Chicago News*, while another fifty were in advanced stages of assembly. Whitelaw Reid was still not satisfied. His insistent messages included specific instructions for changing some of the manufacturing procedures. He hired friends and relatives of friends, then instructed Ottmar to assign them to jobs. In one instance, Reid even rehired an employee whom Ottmar had discharged for incompetence.

The company was soon going to re-capitalize and stockholders were investing additional funds. Mergenthaler owned 1,000 shares of stock in the old company and now found himself seriously embarrassed by the re-investment option. His position as inventor morally compelled him to share the risks of the other stockholders, and his influence in the company was thought to depend largely upon his doing so. Yet he was entirely unable to meet the assessment amount of $25,000; his entire fortune at that time amounting to only about $8,000, and this was locked up in his house and his shop equipment. An appeal to the directors of the newly organized company to accept his subscription and allow him to pay it later out of royalties which would accrue, was refused. Finally Whitelaw Reid volunteered to pay Mergenthaler's assessment, on the condition that he receive the stock as security, together with an irrevocable proxy, and that 6% interest be paid on all money thus advanced. Reluctantly, Ottmar accepted, though it hurt his pride considerably to be thus thrown upon the generosity of Reid who was president and general manager of the new organization, and Mergenthaler's immediate superior.

A great number of special tools and devices were made for the Camden Street factory. The editor of a Baltimore daily paper, after seeing the Linotype and the means employed in producing it, expressed himself to the effect that the ingenuity and energy shown in the organization of the factory were second only to that shown in the construction of the machine itself, and that he could hardly conceive it possible that the two factories in Baltimore should have been created and equipped in so short a time.

One great difficulty in the matrix department was found to be the expense and trouble of maintaining the original steel stamps which produced the matrix proper. There was no machine existing at that time by means of which these stamps could be engraved at a small cost with an absolute certainty of maintaining the same size and shape, and had to be engraved by hand at a cost of $5 per piece, and their accuracy as a matter of course was far inferior to those later produced by the Benton & Waldo type engraver. Without the help of this machine it may well be said that good matrices would have been an impossibility. Ottmar soon discovered the need for such a machine, and failing to find anything on the market, he designed his own engraving machine. The work on this machine had already well advanced when others brought out theirs, thus causing work on his design to be discontinued.

Ottmar, not being satisfied with the foreman in charge of the matrix department, was compelled to discharge him. Reid re-employed the man, raised his wages and established a special shop for him where he was to make matrices in competition with the Baltimore establishment. This activity went on for years and never produced matrices which were used for more than a few days. The dispute probably arose over disagreements on the material and manner used for producing matrices.

In another instance Ottmar had in one of the factories a clerk who showed himself under the influence of liquor and neglected his work. He being a brother of a prominent stockholder, Ottmar politely told him that he had to do better in future. A few days afterwards an order was received from New York for independent tenure of the man under the board and his wages were increased. It was evident that there were growing resentments on both sides.

A major area of friction concerned the matrices. The New York group had refused to accept Ottmar's experiments with steel matrices. In one of the few printed pieces to come out of the

Baltimore plant, it was noted that "[they were] engaged for a number of years in efforts to produce matrices of better wearing qualities than those made of brass . . [and] have solved the problem by using soft steel and hardening the matrix around the impression point." Hand-cut punches were expensive and not indestructible. There were six or seven punchcutters employed who could not keep up with the demand beyond two hundred machines.

With a view of encouraging the more efficient men Ottmar introduced a bonus system, under which every assembler was entitled to a bonus of $10 for every machine satisfactorily assembled and accepted by the company, always provided that the total cost of assembling did not exceed a certain amount. Under this system, the efficiency of the assembling force increased rapidly and the cost of assembling was reduced. One of the men soon managed to assemble two machines per week, thus giving him $20 extra pay besides his ordinary wages. The machines assembled by him cost far less than those assembled by any of the others. The system having proved a success on the assembling floor, Ottmar introduced contract work in the manufacturing department and soon the efficiency of that branch increased and the work was brought out much cheaper without any lowering of quality.

On February 25, Reid, who had already been urging Mergenthaler to avoid beginning new work, and thus to reduce his payroll if possible, wrote to him as follows:

> "After consultation with a considerable majority of the Board of Directors, I have found it the unanimous opinion that no steps should be taken towards the manufacture of a third set of 100 machines in Baltimore until the company are able to see their way more clearly out of what has already been undertaken, the completion of the second 100 and of the 41 still remaining on the first hundred. The reduction of the pay roll, instead of being the calamity you seem to consider it, would be regarded as a real advantage by the Board of Directors, and it is desirable that in doing it you should weed out the least useful men. [Criticism of Mergenthaler's management was supposedly lost.] My earnest advice to you

is to endeavor to reduce the expenditures in all these particulars to the most economical basis."

In his reply, dated February 28, Ottmar said:

"The conclusion arrived at regarding the work in the Baltimore factories must certainly be considered as most distressing and injurious to the interest of the company. If the policy outlined is to be adhered to, it will amount to an almost entire suspension of the Camden street shop and the consequent loss of the most intelligent set of men, which to get together was the work of years. I will continue to fill the position to which I was appointed myself; respectfully declining to follow the kind advice given."

In answer to this, on March 9th, Reid said:

Sir:
    Your recent letter declining to follow the advice given in my last note, "To endeavor to reduce the expenditures to the most economical basis," is at hand.
    You are hereby directed—(1) To suspend any work, if any has been begun, on any part of the third set of one hundred machines.
    (2) In view of your statement that a considerable number of employees would be idle if such work were not begun, you are hereby directed to reduce the force at once to such an extent that no one shall be retained in our service who cannot be profitably employed on the work now in hand in the completion of the first and second sets of one hundred machines each. In executing this order, you will take pains, of course, to retain those operatives whose employment is most useful and economical for the company.
    (3) You will report as early as practicable what reduction in the weekly pay roll can thus be effected.
    (4) You will confine purchases of material for the company strictly to what is still needed for the completion of the machines of the first and second sets of one

hundred each, and will report as early as possible what amount of material you think still likely to be needed for the completion of these machines. Respectfully,

Ottmar Mergenthaler's resignation, dated March 15, 1888:

Dear Sir:

Your orders for the 10th have come to hand. Regarding the supplies still needed for the completion of the work now on hand I estimate the same to cost about $3,500.

Regarding the reduction of the pay roll at the Camden Street Factory, I can say that within three months the same can be decreased to practically nothing; how much such decrease will amount to from week to week I am unable to tell, as the men will have to be dropped just as they get done with the work on hand.

As to the carrying into effect of the order to close the Camden Street Factory, it is with reluctance that I see myself compelled to leave the same to my successor, and herewith tender to you my resignation as Manager of the company; the same to take effect as soon as convenient, certainly, however, before the order referred to has to go into general effect. The reasons which impel me to this step are generally known to the Board and yourself, and have already formed the subject-matter of a number of communications and complaints on my part to that body or yourself as their representative.

I mention letters of May 18, 1886, March 10, 1887, April 25, 1887 (to Mr. Wm. H. Smith), November 1, 1887, November 14, 1887, and November 24, 1887.

When I was first approached on the subject of giving up my well paying machine shop and to work for the company under a salary I was not at all anxious for such a change, and finally only consented on the condition that I was to be sole manager of the shop and judge of all technical questions, a condition without which I did not see that I could successfully carry out the enterprise to which I was to be appointed.

The contract, dated November 13, 1884, was drawn up in that spirit, and contains a clause which, although in different wording, was supposed to cover that point fully.

Under this contract I have since been faithfully working for the best interests of the company, and have complied with all its provisions to the letter. Under this contract the company did get into possession of its most valuable patents which in good faith I have assigned to the company, although I am sorry to record here that on the company's part most of the provisions are still to be carried out, and that provision of my being sole manager and judge of all technical questions has never, since the advent of the Mergenthaler Company, been recognized, and was, therefore, from the start a point of contention between myself and the Board.

I leave it to the Board to decide whether the company was benefited by such disregard of the plain provisions of the contract under which I consented to render my services, but whatever may be their opinion on this point, I cannot consider any longer to hold a position for which I did not engage myself, and which I never intended to fill.

Force of circumstances, particularly the fact that a large number of worthy people have invested in this enterprise, pinning their faith in my ability to carry it to a successful end, have so far prompted me to continue in charge of a work which I had to perform under so entirely different conditions than intended and under so many needless difficulties; but now I feel that I have complied with every promise ever made by me, and that in now leaving a position which I could not fill satisfactorily to myself and the Board both I am performing but a plain duty towards myself and the company.

I leave the company (all reports to the contrary notwithstanding) as a full commercial success in a condition to economically reproduce their machines in its own factories in numbers large enough to supply the present demands, and with a staff of competent men, who, under prudent management, are able to carry on the business of the company without further direct connection with it on my part.

As to my future relations with the company, I wish to say that I will do all in my power to assist the same in any way I can, always of course provided that the company will carry out the provisions of our contract in the same good faith in which I have carried out mine.

In any work which the company may have done outside of their own shop I should like to be considered on the same basis on which others are, but I should never again, under any consideration, consent to further work under a fixed salary, or to exclusively work for any one concern and excluding all other opportunities.

In entering into the contract with the National Typographic Company it has been part consideration on my part, that by so doing it would open up an avenue for the realization of a long-cherished ambition of mine to become the head of one of the largest commercial establishments in Baltimore. The exercise of my ability as an inventor I only regarded as the route to that end.

The action of the Board in suspending work is tearing down the foundations of the structure, which to build was the result of years, and the collapse of which I cannot witness from within. I prefer to pick up the material thus thrown away and use it in the building of a new structure built upon a different basis, feeling that by doing so I can best serve the interest of the company which is bearing my name, the welfare of which I still have as much at heart as ever.

Thanking yourself and the Board of Directors for the personal deference always shown to me, and feeling extremely sorry that our official relations could not be equally harmonious as our personal ones, I remain, Yours truly,

Ott. Mergenthaler

On March 17th, 1888, a letter was received:

Sir:

Your letter of the 15th was duly received, and will

be laid before the Board at its first meeting.

I beg to call your attention to the fact that my letter of the 9th inst. contained 4 specific orders, the purpose of which will be defeated by delay. Please advise me at once whether you are obeying each and all of these orders or not.

Respectfully, Whitelaw Reid

Baltimore, March 19, 1888.

Dear Sir:

Your letter of the 17th just at hand. In reply I should say that my answer of the 15th substantially covers all points made. I have followed orders No. 1, 3 and 4. Regarding No. 2, I thought I made myself plain; that is, I will of course not carry any employees on the pay roll who cannot be profitably employed on the work started, and a number have been dropped already. But, as appears by my former letter, I do not want to remain in my position long enough for the orders to go into general effect.

Yours truly, Ott. Mergenthaler.

New York, March 20, 1888.

Sir:

Your letter of the 19th inst. implies that you are not carrying out the order No. 2 in my letter of March 9th.

In view of your statement that a considerable number of employees would be idle if work were not begun on parts for a third set of 100 machines, that order directed that you should reduce the force at once to such an extent that no one should be retained in your service who cannot be profitably employed on the work now in hand in the completion of the first and second sets of 100 machines each.

You say that you do not want to remain in your position long enough for this order to go into general effect. This order was to go into effect at once. Your wishes as to future service will be laid before the Board of Directors at their meeting. Meantime, however, the object of the order is defeated by delay; and failure on

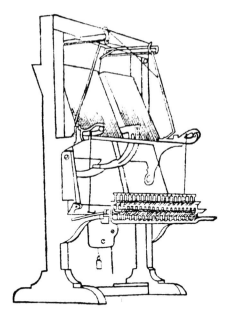

The first composing machine on record was the 1822 device invented by Dr. William Church.

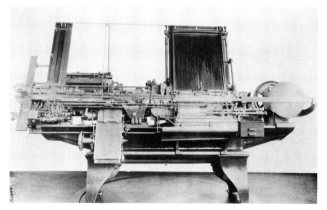

The Paige Compositor was the "dam" machine that Mark Twain subsidized.

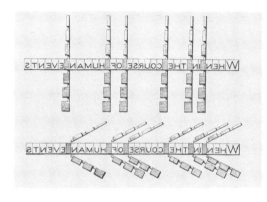

The "step justifier" used fixed widths for justification.

The Brooklyn factory. The insets show the growth of the factory from its beginnings in 1888.

Phillip Tell Dodge, patent attorney and later president of Mergenthaler Lintoype Company.

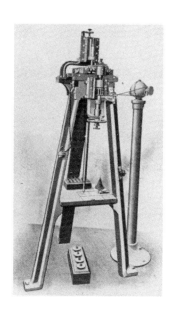
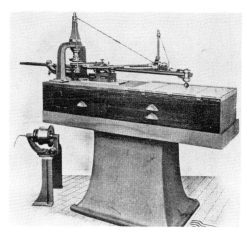

The Benton and Waldo Punch Cutting Machine. The upper left figure is actually the second machine built by Benton and the one on which punches for Mergenthaler were first cut. A later version is shown at upper right. The lower figure shows the various stages of manufacture of the punch (l. to r.): blank steel, blank with shoulders milled out, and completed punch.

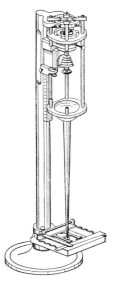

The original Benton and Waldo punch cutter (1885) from the patent.

Linn Boyd Benton. His punch cutter made the Linotype a viable device.

The version of the Blower Linotype which set the New York Tribune, starting July 3, 1886. This machine sat in the Mergenthaler Linotype office until 1967 when it was donated to the Smithsonian. In the 1970s it was refurbished by a British firm to be made operational.

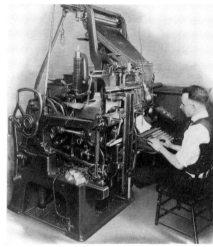

The 1890 Square Base Model 1 Linotype.

The Linotype as it finally appeared.

The slugs were inked and a proof was made by pressing paper against them.

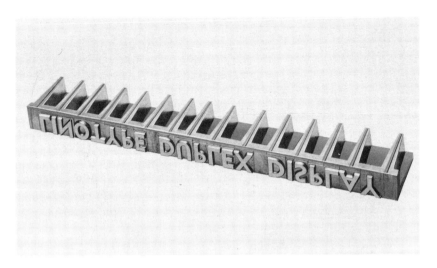

A slug of type. This slug used a recessed mold which reduced the amount of lead required.

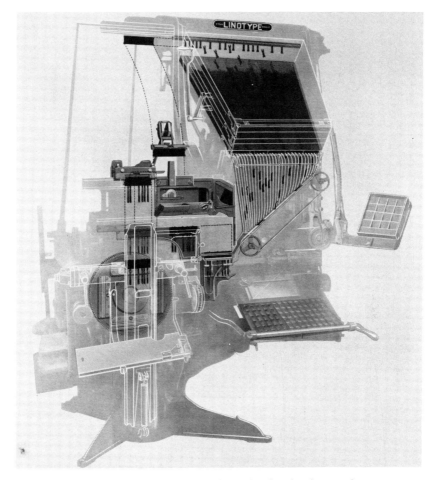

This famous phantom Linotype shows the path taken by the circulating matrix from magazine to assembling elevator to casting to distribution.

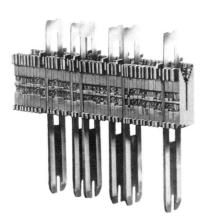

A line of matrices and spacebands (justifying wedges). As the spacebands were pushed up (all at one time), the space between the words expanded, pushing the matrices to the left and right margins, in order to justify the line.

The first matrices had only one character. After 1900, all matrices had two characters, one in one font and the other in another font. Thus you would normally have a type character and its italic or bold version. The number before the triangle told the point size and the number after it designated the type style.

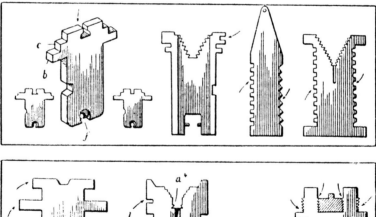

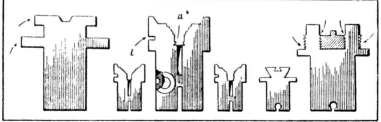

Mergenthaler experimented with many different forms for matrices.

Contemporary illustrations of the first Linotype.

A supposed scene with Whitelaw Reid and Ottmar Mergenthaler on the fateful day of July 3, 1886, when Reid is said to have uttered the famous "line-of-type" name.

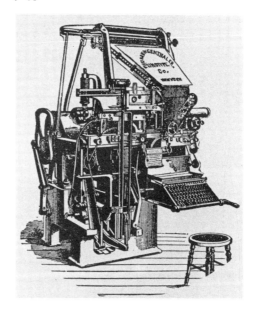

The Simplex Linotype (1890) provided the pattern for all later linecasters.

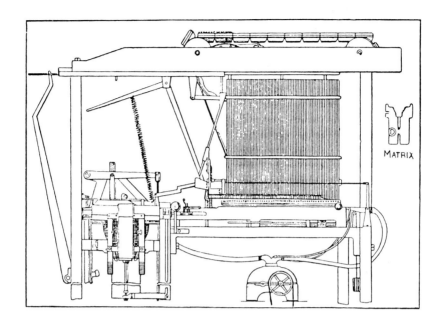

The first Blower machine (1885).

MACHINE WRITING AND TYPESETTING

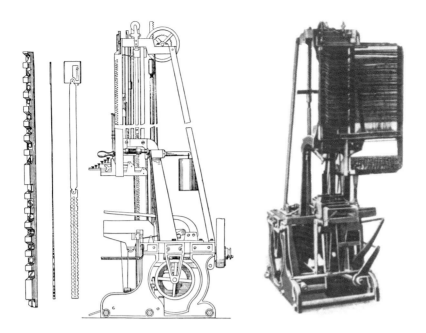

The Direct Casting (or Second) Band Machine was the first to cast a line-of-type. Vertical bars contained an alphabet of female matrices.

The Bar Indenting (or First) Band Machine. Each vertical bar had a raised punch for each letter of the alphabet. After a line was assembled the characters were stamped into papier-mâché for later casting.

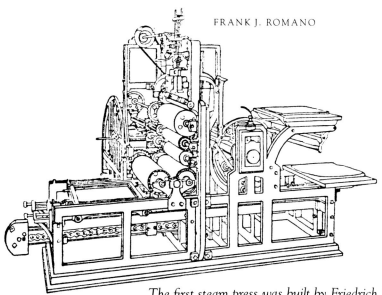

The first steam press was built by Friedrich Koenig in 1812.

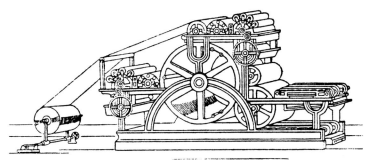

The Bullock press of 1865 was the first to use a continuous reel (or web) of paper.

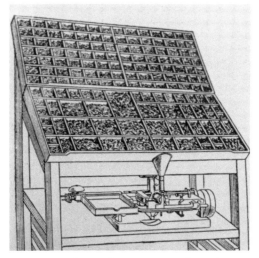

The Typotheter had the compositor throw the type into a funnel.

Charles T. Moore's first attempt at a typesetter used a double cylinder approach.

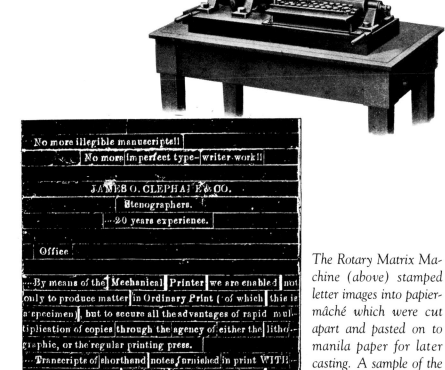

The Rotary Matrix Machine (above) stamped letter images into papier-mâché which were cut apart and pasted on to manila paper for later casting. A sample of the output is shown (left).

Mergenthaler's birthplace in Hachtel, Germany, with memorial plaque.

Monument to Mergenthaler in Loudon Park Cemetery, Baltimore.

A galley of slugs. Corrections to typeset copy had to be made a line at a time.

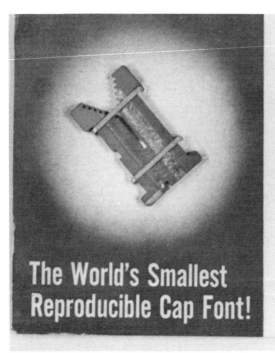

A very rare promotion demonstrating the quality of type cast from the Linotype matrix. The letters A-Z printed out in about one point type.

your part to obey it becomes defiance of the general policy of the Directors, resulting in the waste of the company's money. It is imperative that the order should go into operation at once. A deliberate refusal to obey it could be treated only as a breach of contract, to be met by removal.

Regretting that you should drive me to the necessity, I now enquire for the last time whether you will carry into effect immediately the order No. 2 in my letter of March 9th? At the same time, I have to give notice that liabilities incurred by you in defiance of this order will be your personal liabilities, not those of the company. The fact that this week's pay roll shows practically no diminution makes this notification necessary.

Respectfully, Whitelaw Reid.

Dear Sir:

In answer to your communication of the 20th, please note that my letter of the 19th does not imply any such refusal as pointed out. I said plainly that I would of course not carry any employees on the pay roll which could not be profitably employed on the work now on hand. The payment has been reduced to the extent called for by your order. As to the future, an immediate acceptance of my resignation will protect the Board fully, and there ought not to be any need of removal under the circumstances.

As for the intimated breach of contract, if there is any it is on the other side. I have never consented in my contract with the company to work under direct orders such as they are made now.

Yours respectfully, Ott. Mergenthaler.

From the foregoing it will appear that although Ottmar had resigned, Reid would not accept the resignation but demanded an unconditional compliance with his orders under threat of dismissal.

Simultaneously with his resignation Ottmar had asked a settlement of his account with the company consisting of a claim for his tools amounting to about $6,000 and 10% royalty on the machines built and brought into use to date. Reid replied that the claims would be laid before the Board.

On April 4th Mergenthaler's resignation was accepted and he was asked when it would be convenient for him to turn over the factories to his successor. In reply he suggested the propriety on the part of the company to effect a settlement of his tool claims before they asked him to turn over a property in which he still held a specific part as his own. Reid, however, immediately replied that his tool claim had no connection with his resignation and demanded an immediate surrender of the factories, which order was promptly complied with. Reid assigned crews to transfer the factory at once from Baltimore to a new site on Ryerson Street, in Brooklyn, New York, near the Navy Yard. In the haste of the move many of Ottmar's personal tools and equipment were taken, along with the company's machines. When he protested and demanded Reid return his property or reimburse him, he was ignored.

Ottmar now found himself unemployed, heavily in debt for assessments on his stock subscription, and without means of reestablishing himself in business except by the sale of his share interest in the company. With Reid trying to avoid liability under Mergenthaler's contract with the company, Ottmar sought the legal advice of Charles Marshall, of the Baltimore bar, as his attorney in both the tool claim and the royalty claim.

In answer to a letter by Mr. Marshall, the company declined responsibility under either claim, pretending that the royalty claim depended on contingencies which had not arrived and might never arrive, while in the matter of the tool claim they expressed a willingness to return the tools on proper identification, but declined to be responsible for interest, wear and tear, or loss of any sort. With Ottmar's tools incorporated with the rest of the tools and having been so used for nearly four years, identification and selection of the smaller tools was out of the question.

Under the circumstances Ottmar did not feel like attempting the impossible task of having his tools identified and extracted, and on advice of his attorney brought suit for both tools and royalty. The company was clearly liable under Mergenthaler's contract with the National Typographic Company. Take away that contract and the Mergenthaler Company would not have had the shadow of existence. Reid conveniently got over that point by pretending that his company assumed only the rights accruing under that instrument, but that all obligations carried with it were still to be discharged by the National Typographic Company.

The spring and early summer of 1888 passed and Ottmar was still without tools and without money from the company. The tool question assumed another shape almost every week, with Ottmar receiving notice from the New York office that the tools were ready for delivery under some condition, only to be informed by the superintendent of the factory that he had no such orders or that his orders differed from those stated. Later, through a letter by William Henry Smith, the Secretary of the company, Ottmar was informed that it had been decided to settle his claims and that if he would come over to New York to discuss the details there was no doubt that the matter could now be satisfactorily arranged. This was good news and unexpected at that. What could have brought about such a sudden change of sentiment? Ottmar hastened to New York and to his surprise found that the meeting was not to take place at Reid's office, but at some room in one of the large office buildings. The secretary of the company was there, but almost immediately on Reid entering the room he left, leaving Reid and Mergenthaler alone by themselves.

Reid started out to say that the company had decided to settle the dispute and that it would pay the tool claim in full with interest. "Of course," he continued, "it would be of no use to settle one claim unless by so doing we can settle the other too, and all that we ask you in this respect is to give us a release of the royalty obligation." "So you ask me to abandon all my rights and all the compensation which I have to expect for my valuable invention, in return for having a bill paid which only amounts to a few thousand dollars and which I can collect through the courts, if not otherwise! No, sir, I will never do such a thing; as it is, the courts will have to settle this matter." "The courts!" Reid rejoined, "we have nothing to fear from the courts, it is an unjust contract and it is an illegal contract and courts are not very much given to enforce such contracts as that one." Reid spoke of the earnest desire of the board to arrive at an equitable settlement, and Ottmar asked what he thought an equitable settlement to be. "An equitable settlement in my opinion consists in giving you nothing, at least not on the machines used by the syndicate, but the board may think differently on that point." This was enough for Ottmar; he abruptly ended the meeting, more than ever wondering about Reid's queer ways of assisting him in building castles on the Rhine.

After a decade of labor and devotion to the development of his invention, he was without funds and without even tools and machines to support his wife and four children.

August Hahl offered Ottmar an opportunity to rejoin him. "We were partners once, let us be partners again," he proposed. But Ottmar replied "No, August, I thank you from the bottom of my heart," replied, "but I cannot go back to you, I have brought this trouble on my own neck, and I must rid myself of it through my own efforts."

He began to sell off his shares of stock in the Mergenthaler Printing Company to make a first payment on a new machine shop. He also secured a loan to purchase machines and tools by using as collateral his interest in the equipment which had been moved to Brooklyn. He found himself with some free time which he used to work on drawings for improving the Linotype.

Blower model air blasts made a lot of noise and the mechanism was complex and difficult to repair except by highly trained company mechanics. These were some of the criticisms he had received from Linotype operators. His plan was to eliminate and combine parts, simplifying construction and eliminating some of the sources of mechanical difficulty. He made dozens of changes, including the elimination of the compressed air feature altogether in favor of a redesigned chute assembly.

On July 2, 1888, tragedy struck. Little Julius died. It left Ottmar heartbroken.

By October blueprints for the new model were finished. Although the fundamental principles were exactly the same as in the Blower Linotype, the latest design called for a far simpler and more compact machine which Ottmar named the "Simplex" model.

With the new shop almost ready, he now turned his attention to financing an experimental model. He went to Washington to see Hine and Clephane with his drawings. The Simplex model would not only be cheaper and less bulky, but it would also be more rugged and far less prone to mechanical breakdown, he explained. "Unfortunately, I cannot afford to construct a model. Otherwise I would already be at work on it."

Hine and Clephane promised to see what they could do. They complained about the way things were going in the Mergenthaler Printing Company. Since Ottmar had left and the factory moved to Brooklyn, production had broken down completely. The few machines that had been turned out were so defective as to be completely worthless. Clephane was particularly outspoken in his bitterness toward Whitelaw Reid and the publisher's syndicate. "They have virtually destroyed the manufacturing facilities and are close to permanently killing the Linotype's chance for success. Many of the backers are aroused. If things do not improve, Ottmar, we may soon see a change in the management of the company." Hine agreed with Clephane.

Ottmar's health was another problem and it was learned that he had pleurisy. Ottmar was bedridden for nearly two months and recovery was slow and painful, but at last the cough disappeared and the ache in his lungs diminished.

Clephane showed up in Baltimore. "I want to borrow the drawings for the Simplex model," he said. "Hine and I felt that nothing should be done while you were ill. Now we want to see what we can do to help." The next day, the stenographer was back with an envelope containing checks and cash totaling $2,000. "These represent contributions of two hundred dollars each from ten of us," he said. "I'm sorry that there wasn't time to get to the bank and convert the money into a single draft, but I wanted to get it to you as soon as possible. It need be repaid only if the Mergenthaler Company agrees to construct the Simplex Linotype commercially and compensates you for it."

CHAPTER 9

THE EIGHTH WONDER OF THE WORLD

There were many difficulties which Ottmar had to overcome in manufacture, mainly the rapid production of matrix punches, which had traditionally been cut by hand. The necessary imperfections of such a process led to many problems, particularly unequal depth of punching and the breaking down of the matrix side walls. By one of those curious coincidences of fate, there was a man in Milwaukee who for quite different purposes had experimented with and solved the same problem. This was Linn Boyd Benton who in collaboration with a friend named Robert V. Waldo, had made a punch-cutting machine in 1884 for use in typefounding. It was a pantograph attached to an automatic borer.

This device came to the knowledge of Philip T. Dodge through a review of patent information. He was then the Patent Attorney of the Mergenthaler Company. There was nothing more necessary to ensure the commercial life of the Linotype than this machine, and an arrangement was arranged with Benton and Waldo for the use of their invention. The first punch-cutting machine was rented from them on February 13, 1889 and installed in Brooklyn. Ottmar probably had a machine in Baltimore but no record of its installation exists.

Benton was born at Little Falls, New York, on May 13, 1814. He had started out to invent a typesetting machine but produced a device far more important. Faced with the problem of having to cut some three thousand punches, with no punch-cutters available, Benton designed a machine to perform this laborious task, freeing typefounders of their dependence upon hand-cut punches just in time to aid in supporting their nemesis: machine composition. Based upon the method of the pantograph, the Benton punch-cutter was patented in 1885. Benton died in 1932, fifteen days after he retired. His son, Morris Fuller Benton, joined ATF in 1896 and went on to become one of the most prolific type designers of all time.

Without this device the composing machines being developed during the same period could scarcely have been practical. The machines of Mergenthaler, Lanston and others needed a rapid method of punching matrices in order to sell them in quantity to accompany the machines, and hand-cutting the punches would have been impossibly laborious. The Benton machine was purchased for this purpose by both the Linotype and Monotype firms.

Benton's foundry was merged with the American Type Founders Company in 1892. His machine was later adapted for direct engraving of matrices, the practice followed by the typefounders thereafter.

On January 5, 1889, Mark Twain, once again, waxed eloquent:

> "At this moment I have seen a line of movable type, spaced and justified by machinery! This is the first time in the history of the world that this amazing thing has ever been done."

The "dam typesetter", as he would call it later, was the machine known as the Paige Compositor, the brain-child of an eccentric Rochester, New York inventor named James W. Paige, called by Twain "a most extraordinary compound of business thrift and commercial insanity." The "dam typesetter" cost Twain close to a quarter of a million dollars and was the direct cause of the bankruptcy of his publishing business and personal financial losses that put him on the lecture circuit for years in an attempt to pay his debts.

But he kept pouring money into its development while extolling the virtues of this "miracle". In a letter to his brother, Orion Clemens, he said:

> "This is indeed the first line of movable types that ever

was perfectly spaced and perfectly justified on this earth. This was the last function that remained to be tested—and so by long odds the most important and extraordinary invention ever born of the brain of man stands completed and perfect . . .

"But's a cunning devil, is that machine!—and knows more than any man that ever lived. You shall see. We made a test in this way. We set up a lot of random letters in a stick—three-fourths of a line; then filled out the line with quads representing 14 spaces, each space to be .035" thick. Then we threw aside the quads and put letters into the machine and formed them into 15 two-letter words, leaving the words separated by two-inch vacancies. Then we started up the machine slowly, by hand, and fastened our eyes on the space-selecting pins. The first pinblock projected its third pin as the first word came traveling along the raceway; second block did the same; but the third block projected its second pin.

"Oh, hell! stop the machine—something wrong—it's going to set a .030" space!" General consternation. A foreign substance has got into the spacing plates." This from the head mathematician.

"Paige examined. No—look in, and you can see there's nothing of the kind." Further examination. Now I know what it is—what it must be; one of those plates projects and binds. It's too bad—the first test is a failure." A pause. Well, boys, no use to cry. Get to work—take the machine down.—No—Hold on! don't touch a thing! Go right ahead! We are fools, the machine isn't. The machine knows what it's about. There is a speck of dirt on one of these types and the machine is putting in a thinner space to allow for it!"

"All the other wonderful inventions of the human brain sink pretty nearly into commonplace contrasted with this awful mechanical miracle. Telephones, telegraphs, locomotives, cotton gins, sewing machines, Babbage calculators, Jacquard looms, perfecting presses, Arkwright's frames—all mere toys, simplicities!

The Paige compositor marches alone and far in the lead of human inventions."

Within a few years he would curse this machine for the fortune it wasted. However, the publicity he generated for machine typesetting inspired other would-be inventors to proceed.

Adjustable horseshoes, a mail bag lock, a hydraulic dumb waiter, and an adding machine were all developed and patented by Tolbert Lanston before he began to experiment with a competitive approach to Mergenthaler's Linotype. As a clerk in the Pension Bureau, it is said that he was influenced by Herman Hollerith who used the equivalent of the punch card and tabulating machine to aid in the 1890 Census. In 1885 Lanston applied for a patent which was granted in 1887 and embodied a unique approach to typesetting which is now considered a part of electronic typesetting—the separation of the keyboarding operation from that of actual typesetting.

Lanston's keyboard delivered perforated tapes (the punch card influence?) which contained the necessary information for justification of a line of type in addition to the selection of the characters to be cast. In the original machine type was not cast but was impressed into pieces cut from metal strips. One of the punched paper ribbons set this strip in motion and controlled its movement in accordance with the width of the character to be made. The single type was then cut from the strip automatically and impressed with a female die or matrice containing the desired character. The second punched tape positioned the die-case, which contained 196 matrices.

Lanston's second model cast type from molten metal. He then perfected the keyboard to permit the use of only one punched tape ribbon, which positioned the matrices in the casting machine and delivered characters and spaces that were assembled into justified lines of type. His final patents were granted in 1897.

The development of Lanston's Monotype was placed in the hands of engineer J. Sellers Bancroft. Less than one hundred machines were produced from Bancroft's model, the first being installed in the plant of Gibson Brothers in Washington, DC in 1898. By 1901 there were 94 in use. A few years later Lanston suffered a stroke which incapacitated him until his death in 1913.

Other people would finally complete the Monotype and it would gain fame internationally through the efforts of the British company, a totally separate organization from the American manufacturer, Lanston Monotype.

Meanwhile, back at the Linotype, Clephane's prediction of a stockholders' revolt against Whitelaw Reid's management was accurate. In January, 1889, Ottmar learned that the Washington stockholders had pooled their funds to purchase enough additional shares to wrest control away from the publishers' syndicate. "It will be a bitter fight," Clephane told him. "Reid's group has inexhaustible funds. We don't. Yet we must make the move if the company is to survive."

Reid and the other publishers did not fight back. It was as if they wanted to relinquish control. During two weeks of intensive activity, Hine and Clephane, acting for the Washington group, located and bought up enough stock to obtain a majority interest. During this time, the publishers did not try to stop them.

In January, Whitelaw Reid was deposed as president of the Mergenthaler Printing Company, and Hine was elected to replace him. Hine immediately undertook an investigation of the company's financial status and his discoveries cleared up the mystery of why Reid had refused to fight.

The financial picture was blacker than anyone had pictured. The move from Baltimore to Brooklyn had cost an enormous sum and production at the Brooklyn plant was at a virtual standstill. The machines that had been turned out contained so many mechanical defects that many had to be returned for repair or complete overhauling. Rumors persisted throughout the publishing and printing industries that the Mergenthaler Linotype was a failure. Potential buyers had been frightened off, and those who had already submitted orders canceled. At the time that Hine took office there was not a single order on hand.

The Washington group had to shore up the financial structure of the company, save the reputaton of the Linotype itself and re-establish commercial production on a sound footing. Hine's first move was to call on Ottmar for help. "I intend to cancel production of the Blower Linotype at once," he said. "When your Simplex model is constructed, that is the machine we will make and sell. How soon can you have it for commercial manufacture?"

"I think we can plan on producing the machines for sale by the end of the year," he predicted.

"Not earlier?" Hine inquired.

"Not if the machine is to be mechanically reliable."

"We'll gear our plan to your schedule," Hine promised.

The spring, summer and fall months of 1889 were difficult ones as Ottmar once again worked night and day to rush the new model to completion. Hine and Clephane had worked out a plan to obtain new customers for the Linotype when it was ready for manufacture. Instead of selling the machine outright, they would offer to lease them on a monthly or yearly basis, so that the buyer would have a chance to test them before purchase.

Hine also arranged an agreement with Ottmar to have his Baltimore shop manufacture the machines for the Mergenthaler Company on a fixed contractual basis. "We will pay you a regular sum for each machine you make for us, just as with any outside contractors. That will be in addition to your normal royalty as the inventor of the Linotype. In this way, the company will have the use of your manufacturing facilities as well as its own in Brooklyn. It will benefit the company, and it will benefit you."

In February, 1890, the first Simplex Linotype was shipped to New York City and demonstrated at a special exhibition arranged by Clephane in the Judge Building in Manhattan. Invitations had gone out and Clephane, energetic and persuasive as usual, paid personal visits to convince doubters to attend.

After the first day's demonstration, the success of the exhibit was no longer in doubt. The machine worked so smoothly and efficiently that even the most skeptical conceded that they were impressed. Attendance at the exhibit multiplied, and the guest book soon included the names of almost every printer and publisher in New York City as well as visitors from other major printing centers.

The Mergenthaler Company's offer to lease the machines constituted a persuasive marketing tool. Any legitimate printer could now try out the Linotype in his own composing room without risking financial loss other than the nominal rental fee. By the close of the exhibit, Clephane reported that nearly a hundred orders for the leasing or purchase of machines were already on file. And within two months the number increased to more than four hundred.

The introduction of the Simplex Linotype ushered a new era in printing. In less than a year, Linotypes were being installed in newspaper composing rooms and printing shops throughout the United States, Canada and England. In a matter of months after the introduction of the Linotype the printing industry experienced a boom unmatched in its history. Newspapers increased their size from four and eight pages to sixteen and even thirty two. Magazines grew to size and frequency and book publishing underwent an unprecedented growth. The cost of printing dropped sharply.

In his second matrix-band machine, Mergenthaler had justified by inserting small wedges between the words, according to an invention of Merritt Gally, better known as the inventor of the Universal Press. Later on, however, J.W. Schuckers, a mechanic at the HomerLee Bank Note Company, of New York, had invented the double-wedge method of justification, using two separate wedges, and this had been acquired by the Rogers Typograph Company. There was a great amount of cross litigation, and eventually the Mergenthaler Linotype Company won against the Rogers machine, but lost in respect of their imitation of the Schuckers' spaceband. Mergenthaler had joined the two wedges in the way that eventually proved successful, but it was decided by the Courts that the Schuckers' Patent had definite priority.

The double-wedge spaceband is one of those mechanisms so simple yet so perfect for its purpose that no other alternative method can be readily conceived. Even the method that was used on several hundred machines, the so-called "step spacer," was not really effective. The difficulty was eventually solved in 1895 by the Mergenthaler Linotype Company purchasing the Rogers Typograph Company, and the Schuckers' patent, for $416,000, while at the same time J.R. Rogers entered the Linotype organization, in which he remained a prominent figure until his death in February 1934.

Until it expired in 1912, the Schuckers' Patent was one of the most important assets of the Mergenthaler company. The legal stand-off between the two companies—Typograph had the Schuckers' Patent but they were enjoined from selling the machine on which it ran, and Mergenthaler had the machine but needed the expandable justifier. It was a brilliant stroke that resulted in the elimination of a potential competitor and the acquisition of an essential mechanism. The Typograph machine was manufactured in Canada and Europe until World War II (and from 1907, when certain Linotype patents expired, until 1912 in the U.S.). In 1960, the German version was re-introduced just in time to be obsoleted by phototypesetting.

The National Typographic Company was the owner of the basic patents and the trading rights for the world. It had about fifty shareholders, and the Mergenthaler Linotype Company of New Jersey had about 333 shareholders. At this time the English Patent rights were transferred to the English Company and the rights for America (excluding Canada) to the Mergenthaler Linotype Company, and it was resolved to end the Mergenthaler Printing Company. The National Typographic Company therefore increased its capital to $2,000,000 and the extra $1,000,000 was paid to the Stockholders in the Mergenthaler Printing Company. Eventually, on March 23, 1893, the Mergenthaler Printing Company was finally dissolved.

The Washington stockholders had taken up Reid's cry of "too much royalty" and a committee was appointed to see whether a modification of the original agreement could not be effected with Mergenthaler. The committee consisted of three of his most intimate friends, men who enjoyed his confidence and they succeeded in doing what Reid could never have accomplished, basing their case entirely on the expediency of Mergenthaler accepting a reduction of royalty in order to re-establish good feeling within the company and induce new faith in the future of the enterprise.

They pointed out to him, that whatever was the underlying cause, the fact remained that so far the enterprise had resulted disastrously to the stockholders and that the prospect before them was still nothing better than a long period of assessments with no show of dividends in the near future. Reid, by his agitation on the subject, had so badly scared the stockholders about the ruinous effect of the ten per cent royalty to the inventor that the interests of all concerned demanded that something should be done to re-establish confidence in the future prospects of the company, and they offered a royalty of $50 per machine, which amounted to a reduction of over one-half of the royalty payable under the ten per cent clause. It was not the intention of the company, they explained, to actually reduce the prospective profits to the inventor, but only to put them into a shape in which they could not be used in future for the purpose of discrediting the prospects of the company, and in return for Mergenthaler's concession they promised that he would be favored otherwise in the line of contracts for the company and increased influence within it.

Hine expresssed it this way: "If Mr. Mergenthaler will consent to this modification he will so endear himself to the company and the stockholders at large that no request of his could be refused, and his position within the company will become impregnable for all time to come, the indirect advantages of which will be worth more to him in dollars and cents than the money he will give up under the proposed agreement."

Mergenthaler yielded, and here he made the financial mistake of his life. In later years, the company would report the $1,500,000 that Mergenthaler's heirs had received up to the point when the patents expired. It was an issue that put company officials on the defensive. Yet, without the Linotype there would have been no company, no stockholders and no profits. Even as late as 1936, in an unpublished company history written by president Joseph T. Mackey, the royalty was called a "curious arrangement".

By the end of 1891, less than a year after the first Simplex Linotype had been installed commercially, the demand for machines was so heavy that the combined production of the Brooklyn plant and Ottmar's own factory in Baltimore could not keep up. A major expansion of manufacturing facilities was required. To obtain funds, Hine mapped a reorganization of the company and new stock was issued, increasing capitalization to five million dollars. At the same time, the Mergenthaler Printing Company was renamed the Mergenthaler Linotype Company. Hine, who had been forced to neglect his own law practice while re-establishing the corporation on a sound footing, now stepped down as president and general manager. His place was taken by Philip Tell Dodge, a Washingtonian who had been serving as the company's patent attorney.

Dodge was born in Fond du Lac, Wisconsin on July 11, 1851. His father later published a newspaper in Minnesota but, at the age of ten, moved with his family to New York state. Like Ottmar, he worked in a machine shop and studied mechanical drawing at night. At the age of twelve, he supposedly demonstrated his abilities to the shop foreman by pointing out that a particular machine had seven unnecessary parts.

He moved to Washington, DC and joined a firm of patent attorneys. He graduated in 1873 at the age of twenty two from George Washington University with a degree in law. His knowledge of mechanics helped him to become a successful patent attorney. He was one of the first people to use the telephone; he bought and used the second bicycle in the Unites States, and also had a hand in improving the typewriter.

Dodge was the patent counsel for Remington and also a user of the typewriter. One of the first patents for the typewriter was typed by him and submitted to the Patent Office. It was returned because the rules stated that patent applications had to be handwritten. Dodge copied the material in longhand.

It was Dodge who secured the lowercase patent for Remington.

Competitive typesetting machines were now coming out of the woodwork. A well organized attempt was made to introduce a "barefaced imitation" of the Linotype, as Ottmar described it, called the Rogers Typograph. This machine was considerably smaller in size and being advertised with as much energy and hornblowing as Barnum's Circus or Higgins German Laundry Soap and looked for awhile as if it would conquer the field before the Linotype interests could carry their case into court and get a decision for infringement. Several of these machines had been placed in the composing room of the *New York World*, with the assistance of G.W. Turner, business manager of the *World* who also had a financial interest in the Rogers machine. The Mergenthaler Company brought suit for a temporary injunction. Their case was handled by the law firm of Betts, Atterbury, Hyde and Betts of New York, assisted by Hine's energy and experience. A preliminary injunction against the Rogers Typograph was issued on March 11, 1891, which injunction sustained and made permanent later, the final appeal of the Rogers Typograph Company being decided against them in December, 1894. As stated before, Mergenthaler then acquired the Roger's company.

About this time also, W.S. Scudder invented the Monoline machine, which he showed in 1893 at the Chicago World's Fair. In subsequent litigation the new Monoline Company was also prevented from manufacturing in the United States.

In April 1890, a contract was given to Ottmar for the construction of one hundred machines at a cost of $1,200 per machine, all complete except matrices and spaces. An equal number was started at the company's works in Brooklyn and work was pushed in both establishments with all energy possible.

The American machine was born in Baltimore and first manufactured in Baltimore. About 1888 the first factory in Brooklyn was started. That factory occupied an area of about one-half an acre, and employed 200 people. By 1936 the factory covered over eighteen acres of floor space, in massive buildings on Ryerson Street near the Navy Yard, employing over 3,000 people. The factory was a monument of specialization, many of the machine tools having been invented by Ottmar or employees of the company.

The matrix-making department was highly specialized, there being no fewer than 49 separate operations before the matrix was ready for use. Over 100 million matrices in over 50 languages were stocked at the height of the Linotype's popularity.

Up to that time nearly two and one-half million dollars had been spent on the machine without one cent of profit, and even after this it was frequently necessary to raise the weekly payroll by loans on the Directors' security. Gradually Dodge's control, and the many improvements in the machine which he either adopted or personally invented, had their effect, and a firm financial basis was arrived at. It has been said that but for the fortitude and vision of Dodge the machine might never have become commercially successful.

It was obvious at this time that still more capital was needed to replace the money which had already been lost. In 1891, the Mergenthaler Printing Company was taken over by the National Typographic Company, and the Mergenthaler Linotype Company of New Jersey was registered on October 28, 1891, in New Jersey, with a capital of five million dollars, $3,000,000 of which went to the Stockholders of the Mergenthaler Printing Company and the National Typographic Company. The new company commenced active business on November 25, 1891, and took over the American property, rights and privileges of the National Typographic Company and the Mergenthaler Printing Company. Dodge became President, Ogden Mills was First Vice-President, Norman Dodge Second Vice-President, F.J. Warburton Secretary and Treasurer, and J. Willis Heard Assistant Secretary and Treasurer, while among the Directors were D.O. Mills, W.C. Whitney, G.L. Bradley, J.O. Clephane, S.M. Bryan, and T.J. Regan. So commenced the successful Linotype company.

There were still areas of friction between Ottmar and the company that bore his name. Ottmar had presented a charge of $1,300 for patent drawings for designing the last machine. The bill was for the work done by him, his draftsman and a patent draftsman after he left the employment of the company in 1888; over half of it represented cash paid to the draftsman and the rest represented Mergenthaler's own time for the period between April and December, 1888, when the Washington stockholders furnished two thousand dollars to carry on the construction work.

During the period covered by the bill Ottmar received neither a salary nor royalty from the company. It had not been presented before for the reason that the charge stood on the books as a charge against Mergenthaler, there being no order for this work from the company for reasons well known to Dodge. Had not the work charged under this bill already resulted in giving the company a yearly income of $100,000? Were there not millions in immediate sight as its direct result? And would not the company have lost at least eight months of time, had Ottmar not used his time and money in the interest of the company at a time when the latter tried to ruin him financially?

To Mergenthaler's surprise, Dodge explained that he could not approve the bill in question because it was not a legal charge against the company. Ottmar maintained the equity of the claim against the company and presented the bill once or twice every year. Finally in 1896, when the company had already made millions out of the enterprise, Dodge offered to pay half of the amount. Mergenthaler insisted on either collecting the whole bill or nothing at all and finally the bill was paid as it stood.

Early in 1893, when the Shuckers' justifier patents had become more and more threatening and Mergenthaler's efforts to invent another practical method of justification had not been crowned with success. The president of the company wrote:

Dear Sir:

New justification.—There is considerable anxiety among the company officers and others as to whether you can overcome the Shuckers justification, and if so, how soon. They are particularly anxious to know how soon the present machine will be completed as to justifier, and whether you consider that it will be a reliable substitute which can be introduced into our existing machines.

In view of the fact that the justifying devices are an essential and necessary part of the machine, without which it would be inoperative, they of course feel that it was one of the essential things on which the royalty of $50 was based, and that if they have got to put something else in order to make the machine operative at all this fact should be taken into consideration as to payments made you for royalty.

Yours truly, Phil. T. Dodge, Prest. and Gen. Mangr.

"Another of Reid's contributions to the castle of the Rhine," Mergenthaler thought and without much delay informed the company that fortunately his relations with the company were not depending on the "opinions and sentiments of its officers," but were regulated by a written contract, the provisions of which he intended to enforce. In answer to this Dodge replied as follows:

Dear Sir:

Sorts Attachment.—I note that this apparatus has been sent to the Brooklyn factory.

Justification.—I note your remarks on this subject; also the matter of contracts. I also appreciate that the company paid you the large amounts, and contracted to pay the royalty at a time when it was supposed that you were giving them a complete machine which they would have a right to make, and not one which would be an infringement of other patents, or which could not be built without being amenable to other people. I note your remarks that the company has not protected you in your creations with patents. The company has fully protected whatever you have completed and done of value. It has not protected you in things which were invented and patented by others and judicially decided to be the property of others.

For a period of eleven months no royalty was paid and every demand for payment was ignored by the company.

In January, 1894, a method of justification by step justifiers had been devised by Ottmar which led to the construction of 225 machines which were all made in Baltimore.

In the meantime the Monoline had made considerable progress. Mergenthaler himself, as early as the fall of 1891 had invented a machine which proved to be practically the same machine as the Monoline, and when his application reached the Patent Office an interference was declared between his own and the application of its inventor, Mr. Scudder. The first hearing in the case took place in the fall of 1893 and at the conclusion of the testimony Dodge expressed to Mergenthaler: "The only only thing against us in this case is the fact that we did not develop our invention while the other parties did. It has become more and more the practice of the Patent Office to award an invention in controversy to the party which reduces its invention to practice."

In June, 1894, Mergenthaler brought out the two-line letter matrix for use in advertisement production, which advanced the usefulness of the machine and overcame one of the objections of newspaper users.

The following correspondence will explain itself:

New York, October 19, 1895
Dear Mr. Mergenthaler:
In considering the new organization it is practically necessary for us to give the new company a new name. It has been found that the present name is so very long as to cause us much inconvenience. We find almost daily that the first part of the name is misspelt in checks, drafts and other papers. Again, the writing of this name, as frequently happens four or five hundred times a day by one person, involves a serious amount of labor. All things considered, it is thought best to call the new

company simply the Linotype Company, but of course we do not feel like making any changes without consulting you in the matter. There is no desire on our part to detract from the credit which belongs to you, or to lessen the publicity which is given your name. Of course in all our headings and advertising matter, and everywhere else, we shall continue to use your name just as we are already doing. I write you at this time, and before any final action is taken in the matter, so that you may not misunderstand our motives, or may not think that we are in any way trying to reflect on you or to detract from the credit and honor which belongs to you.

Yours truly, P. T. Dodge, G.H.S.

October 23, 1895

Dear Mr. Dodge:

Of the many communications received from yourself or from the company you represent, I do not remember one which had made so painful an impression upon me as the one contained in your letter of the 19th, advising me of the intention to strike my name from the official title of the proposed new company. The reasons advanced are, putting it mildly, whimsical. To deprive a man who has given to the world one of the most important inventions of the age of the credit therefor by discontinuing his name, seems to me to be an act unworthy of the stockholders, who have been so greatly benefited by my labors, and doubly so if that act comes coincident to the doubling of the capital stock of the company. From an original investment of no more than one and one-half millions, the company has prospered until now it is proposed to pay interest on ten millions, and on the eve of this event, and as a fitting reward for my labors, you propose to strike my name from the title of the company, for no other reasons than that it is sometimes misspelled and that it is rather long.

It would be interesting to know how much the company is likely to be damaged by reason of the misspelling, and how much clerk time it could dispense with if the name be discontinued.

The company has borne my name during periods when it brought me more ridicule than honor and more aspersions than credit. It has borne it later, and has demonstrated that its name is not inimical to its success. To strike it off now will be a serious and ill-deserved blow and reflection upon me, and the pain caused thereby will be mitigated but little by your assurance that the blow is not intended to hurt me, but that you have to strike me down simply because I am in the way of your catching the fly on the wall.

After all, it seems to be a doubtful policy for a company to materially change its name, under which it has been known and advertised all over the country for the last ten years, and under which it received quite a number of important court decisions.

That the name "Mergenthaler" at the present time is not an obscure and unfamiliar one, but to the contrary, is known to every publisher and printer all over the country, is best shown by the frequent references in trade journals and newspapers in which the machines are referred to quite as frequently as "Mergenthalers" than as Linotypes.

In conclusion I beg to submit for consideration the fact that the exploiters of every other important invention are honoring the inventor by giving his name to their company. Take the Bell Co., the different Edison inventions, the Westinghouse, the McCormick and others, they all carry the inventor's name.

Hoping that I may be spared the intended humiliation, I am,

Yours truly, Ott. Mergenthaler.

The name was not changed until 1985 when it became Allied Linotype, (Allied had acquired Mergenthaler Linotype as part of its acquisition of ELTRA Corporation, in 1979, which owned Mergenthaler) and later with the merger of Allied Corporation and The Signal Companies, it became just plain "Linotype Company". Other than the American company, no other division or international affiliate used the name Mergenthaler.

CHAPTER 10

THE SON WHO DID NOT BECOME A TEACHER

Theodore L. De Vinne, the most famous printer of the time, was asked to state his views on changing technology in 1889 as critics were calling machine composition a destructive force which endangered the future of the industry. He wrote with his usual long-term view:

> Much of the present disquietude is unnecessary. That typesetting by machine may or will reduce the cost of work on reprints and cheap books and papers is probable. That it will ever drive any large body of good workmen out of business is absurd. The machines will surely make more work for workmen. So far from decreasing the standard of workmanship, they will elevate it. In this country there has never been any active hostility to new machinery in the printing business. There have been no moves or strikes against inventions, but workmen look on all new devices with suspicion and unfriendliness. They do not see that the invention which temporarily throws one man out of work, ultimately, makes work for two or more men.
>
> What would be the state of the trade if we had no stereotype or electrotype, no composition rollers, and no printing machines: the daily newspaper as we now have it would be an impossibility. An edition of two thousand or twenty-five hundred copies of a small sheet would be the highest performance of the hand press, and what severe work this paltry performance would impose on the wretched hand pressman who had to

print this edition in a hurry! The illustrated magazine of large edition and low price, filled with fine wood cuts, could not exist at all in the days of hand presses. One could go on and show how hand presses would curtail the production not only of the popular but of the artistic forms of typography.

Processes and machines that were once dreaded are now used by every printer, and they are welcomed as much by the journeyman as the master. No one will pretend that they have reduced the number of workmen. Where there was one printer fifty years ago, there are at least twenty printers now. As a rule, the average piece compositor is a better educated man than the average pressman. Under equal conditions he should and would earn higher wages, but his superior intelligence and education do not increase his production. This production is limited by the slowness of his hand, which is now as it was fifty years ago. If the compositor was employed on a typesetting machine, he would get some of the benefits of the increased production.

One reason why the modern pressman is better paid than the old pressman is because he is a better workman. The machine is more complex than the hand press, and it compels the pressman to exercise more forethought and intelligence. As a rule, the mechanics who bitterly decry machines are those who have been found incompetent to handle them. The men who refuse to learn the theory or the practice of new processes—who are content to do work as it was done when they were boys—who don't want to be bothered' by the study of new problems in handicraft—who evade or shirk responsibilities—are the very men that employers do not want to employ upon their machines. That they may and probably will suffer for their persistent refusal to adapt themselves to changed conditions is much to be regretted; but are they blameless? Is it the fault of the master, or the machine, or the workman himself?"

It is difficult to estimate the extent of unemployment caused by the Linotype. Between 10,000 and 36,000 hand compositors were displaced, but the depression of 1893 – 1897 may have had a major role in that displacement. By 1903, the Linotype and its competitors were creating more jobs than they lost as other parts of the printing process grew as a result of expansion of the publishing industry as a whole. There were three and a half times more publications than existed before the Linotype and newspapers that installed Linotypes had grown so much in the period between 1892 and 1900 that they were employing the same number of people in composing as they had before the introduction of machine typesetting. By 1900, technological unemployment in the printing industry was not a factor.

No group was more vulnerable to machine composition than typefounders. From the establishment in 1796 of the earliest American typefoundry, Binny & Ronaldson in Philadelphia, competition among typefounders had grown to over twenty four firms. Price cutting became ruinous, reaching its peak in the late eighties, as middlemen, or jobbers, who sold directly to the printer were increasing in number and pressured typefoundries to an extent that production standards deteriorated, placing foundries in an unstable financial position. Discounts of 25% to 40% were routine and runious. There were many devices constructed for the composition of printer's types, all of which had employed foundry type, but the Linotype and its competitors cast their own type. Casting machines were immediately successful in newspaper offices, a principal source of income to the foundries. One city newspaper could keep one foundry busy. Thus, each casting machine location literally wiped out a foundry. To survive, the typefounders had to cooperate with one another.

The first overture was made by John Marder of the Chicago Type Foundry with Arthur Brower of the Union Type Foundry, of Chicago. Marder and Brower persuaded Thomas B. MacKellar, head of the Philadelphia's MacKellar, Smiths & Jordan, to join in the formation of a combined firm. Many of the twenty three foundries which joined as the American Type Founders in 1892 could not rid themselves of their old competitive urges and insisted on keeping their original names.

Four large foundries did not join ATF and attempted to rally printers but it was a lost cause, when in 1896, the Lanston Monotype machine was introduced, enabling printers to cast their own types and freeing them from dependence upon the foundries.

For Ottmar, 1892 was a year of triumph—the culmination of sixteen years of continuous effort and dedication to his ideas. With more than a thousand Linotype machines in use throughout the English-speaking world and plans were made to introduce foreign language typefaces so that the invention could be utilized around the world, the Linotype's place in history was secure.

Even Dodge was happy. His annual report for 1892 spoke of increased sales, noting "Happily, however, for the reputation and success of the Linotype the demand for supplies to cover breakage and wear is very rapidly decreasing."

Since his near-fatal attack of pleurisy in the fall of 1888, Ottmar's health had grown steadily worse. He had driven himself so tirelessly over the years that it had become a habit which he found hard to break. At thirty eight, he was beginning to look a man of sixty. He had lost so much weight that he appeared almost emaciated.

That spring, Ottmar's physician insisted that he take a complete rest. Since he had never had a vacation, and now that the Linotype was a commercial success he could well afford one, he and Emma decided to go to Europe. It had been twenty years since Ottmar had last seen his family in Germany. So, in May, the Mergenthaler family boarded a steamer for England. When they arrived in London, Ottmar was astonished to learn how famous he had become. He visited Manchester where a Mergenthaler Linotype factory had recently been established.

He was universally toasted and praised. In Germany he was greeted as a conquering hero. Newspapers announced his arrival, and editorials acknowledged him as one of the leading inventors of the age. In Ensingen, he was reunited with his family. His father, now seventy two, and his stepmother, Caroline, greeted him. Ottmar's brothers and sister and their families came to Ensingen for the occasion, as did his stepuncle Louis Hahl.

The homecoming of its most famous son was a matter of celebration for the entire town and the leading citizens officially welcomed Ottmar to the place he had left as a boy of eighteen. The burgomaster spoke and then introduced Ottmar's father, who said of his son:

"There is little that can be said or needs to be said on this happy occasion," said the schoolmaster. "My son left this village a boy and returns to us a famous inventor. I wanted him to become a schoolteacher, as I and my older sons were, for I saw in the education of the young a worthy calling. Yet in youthful wisdom he saw far more clearly than I that his was meant to be a far more auspicious destiny; in fulfilling that destiny he has done more for the education of mankind that I could ever have dreamed. For me this is the proudest moment of my life."

They returned to Baltimore in the fall. Ottmar immediately resumed his duties in the factory but in a few months the spells of exhaustion and weariness returned. Specialists agreed on a diagnosis. Ottmar had tuberculosis.

On October 27, 1894, while at a health resort, he received a telegraph message that Emma had given birth to a daughter, Pauline. At the end of six months physicians discovered that his conditon had not improved, they arranged to send him to Saranac Lake, New York, where a famous tuberculosis sanitarium had been established. Ottmar lived in Baker Cottage, once occupied by poet Robert Louis Stevenson.

During the months at Saranac Lake he received regular reports from Carl Muehleisen, superintendent of his Baltimore factory. Business was prospering and Clephane wrote to him often with interesting bits of gossip about the Mergenthaler Linotype Company. Orders for machines and matrices were continuing to increase so rapidly that in spite of its recent expansion the company was still finding it difficult to keep up the demand. Moreover, Linotypes were now in use in non-English-speaking countries, a new German Mergenthaler factory having just been established in Berlin.

In the spring of 1896 he left Saranac Lake and traveled alone to Prescott, Arizona. For six months he and a local guide lived in a tent, cooking their own meals, and searching for a climate which would help ease the pain in his chest. In Deming, New Mexico, he found an ideal place where the air was dry and there was a semblance of decent civilization to which he could bring his family. Ottmar rented a cottage and sent for Emma and the children. With them came a young Baltimore native, Carl Otto Schoenrich, the children's tutor and the son of a professor at Baltimore City College. Schoenrich suggested that Ottmar write his autobiography and for many months he and young Schoenrich worked on the manuscripts.

By the fall of 1897 the autobiography was completed, and Ottmar and the youthful tutor began to give it a final reading prior to sending it off to a Baltimore publisher. That evening the Mergenthaler family was awakened by shouts and the sound of people rushing about frantically. A prairie fire was forcing that part of town to be evacuated and Ottmar and the family ran with all the others. The wind had grown stronger and was shifting direction and several houses on the edge of town were directly in the path of the fire. Ottmar remembered that his manuscript was still in the house.

The next day Ottmar and his family rushed back to see their house leveled to the ground. Shortly afterward, they decided to return to Baltimore.

Back home, Ottmar and Schoenrich went to work at once trying to rewrite the manuscript but after a few weeks, Ottmar realized that it was a losing proposition. Holding a pencil became an ordeal and dictating to Schnoenrich for any length of time left him completely sapped of strength. "I am afraid we will have to give up the idea of an autobiography," he said. "Since you helped me with the earlier material and have access to my papers and records at the factory, I would like you to write my story under your own name."

The biography was published toward the end of 1898. Ottmar sent copies to Clephane, Hine, August Hahl and all his old friends, as well as to his family in Europe.

During the early years of the Linotype it was often said that there would be no serious competitor in the composing machine field until it was possible and profitable for some outsider to build a Linotype. One did begin its career in 1911, when Hermann Ridder, of New York, who had always been interested in composing machines, and had been involved with the "Monoline" and "Victorline" machines, established in Brooklyn the Amalgamated Typesetting Machine Co., for the purpose of manufacturing a machine which he was intended to call the "Amalgamatype." The idea of this machine was the casting of individual type, held together in assembly by a metal band. This idea was abandoned, and Ridder entered into association with W.S. Scudder, who had worked for Ottmar in 1887 and had been a member of the Linotype organization and had also invented the "Monoline" machine.

In 1912 the International Typesetting Machine Company was formed, with a factory in Brooklyn, to manufacture matrices and parts for Linotypes and to overhaul and reconstruct secondhand Linotypes. Since the basic Linotype patents had expired in 1909 (and most were void by 1912), it was possible to make a machine like a Linotype. In 1913 the name of the Company was changed to The Intertype Typesetting Machine Company, and the first machine called "Intertype", which was similiar to the Model 5 Linotype (especially since one of its engineers had joined Scudder), was completed. The first unit was installed at the *Journal of Commerce* in March 1913.

Their bank foreclosed on a mortgage of $1,000,000 and a Receiver was appointed who carried on for some time, but in January, 1916, the Company was wound up and its assets were sold by auction. The capital of the Company at this time was $3,977,300—the Mortgage, Gold Notes, and Loans amounted to $1,885,490. There was a suggestion that the Mergenthaler Linotype Company should buy up the concern, but eventually it was sold to a new combination for $1,650,000, and came to life again as the Intertype Corporation with a capital of $2,650,000.

Also during this period, the Linograph was introduced by Hans Petersen. It proved moderately popular until 1944 when it and its company were acquired by Intertype.

For the next 53 years, typesetting users would purchase either a black Linotype or a grey Intertype. Most parts and all matrices were interchangeable. It is interesting to note that during the life of the linecasters, there would be much improvement but no real innovation. The Elektron Linotype, introduced in 1964, was designed to run at fifteen newspaper lines per minute from coded paper tape by elimination of the assembling elevator. It would be a whole new technology—phototypesetting—that would spur the next wave of the "revolution" in typesetting and print communication.

On October 28, 1899, at their home at 159 West Lanvale Street, Baltimore, Emma entered Ottmar's room and called softly to him, but he did not stir. He was dead at the age of forty five.

A few days later he was buried in Baltimore's Loudon Park Cemetery, after quiet services in Old Zion Lutheran Church which he and Emma had attended throughout their married life. He died too soon to witness the full impact of the remarkable invention he had perfected or to see himself honored as one of the inventive geniuses of his age. Thomas Alva Edison referred to the Linotype as the "Eighth Wonder of the World."

Fast, low-cost typesetting, made possible through Mergenthaler's mechanical genius, created a revolution in publishing and education. Until the development of the Linotype, newspapers and magazines were few, thin and costly. Schoolbooks were prohibitively expensive and had to be handed down from generation to generation. Even as late as the 1880's, only seventy six public libraries in the United States contained more than three hundred books. As a direct result of Mergenthaler's invention, the horizons of the publishing industry expanded significantly. Libraries multiplied in number and size and the nation's illiteracy rate was decreased by more than two-thirds.

In 1888 there were 60 Linotypes in use, jumping to 200 in 1889. By 1891 the company was shipping 69 a year, increasing each year: 288 in 1892, 568 in 1893, 890 in 1894, and 1,076 in 1895. By 1904 there were 8,000 machines in use in the United States alone. By 1915 over 33,000 Linotype machines had been shipped, and by 1936 over 50,000 had been made and delivered by the Mergenthaler Linotype Company. It has been estimated that almost 100,000 units were shipped before the company stopped manufacture—1968 in the United States and about 1970 elsewhere. Between the Linotype and all competitive linecasters, the total is estimated to be not more than 160,000 units.

The influence of the Linotype was not confined to America alone. Slugs of metal type were cast in more than a thousand different languages, and Linotypes were used in almost every country in the world.

In 1889, the *Tribune* reported that the Linotype saved them $80,000 in that year alone. In 1904, the New York Press Association held a meeting which discussed the economic benefits of the Linotype. An investment of $3,000, or $10 a week for the machine and $10 a week for the operator, resulted in one operator doing the work of four hand compositors. It was difficult to ignore cost justification like that.

Ottmar Mergenthaler, like Christopher Latham Sholes, cared little for riches or personal glory. Both were modest men motivated by a sense of history and dedication to a dream. Their work initiated the transition of the dream into reality, but many other hands and minds built upon their efforts. Sholes and Mergenthaler launched a revolution that touches all who read.

Reid went on to public service, serving as minister to France (1889–1892) and ambassador to England from 1905 until his death in 1912. His family continued its involvement with the company. His son-in-law, D. O. Mills, was an investor, and his son, Ogden Mills, was a director until 1928, and his son, Ogden L. Mills, was involved in various capacities until 1933.

Dodge continued as president of the company until he retired in 1928, although he was more involved with outside interests after 1912. He served on the Board of Directors of a number of companies, including Royal Typewriter, Columbia Graphophone and International Paper. He was eighty years old when he died in August, 1931.

Clephane did not get exactly what he wanted during his lifetime. It would be almost one hundred years before the gap would be bridged between the typewriter and the printed page. He died in November, 1910, actively involved in promoting the typewriter and the Linotype up until end, strutting about in his high heels and probably looking for new Alps to ignore.

CHRONOLOGY

▶1852
"Mechanical Typographer," J.M. Jones.

▶1854
Ottmar Mergenthaler is born in Hachtel near Bad Mergentheim, as the fourth child of the village schoolmaster Johann Georg Mergenthaler (1820-1893) and his wife Rosine Ackermann (1828-1859).

▶1856
"Printing Instrument for the Blind"; typebars converge at center, Alfred Ely Beach.
    Typewheel principle, J.H. Cooper.

▶1857
"Printing Machine"; four rows of keys in peg form; inked ribbon, Dr. Samuel William Francis.

▶1863
"Improving Mechanical Typographer or Printing Apparatus," F.A. deMay.
    "Improved Hand Printing Device or Mechanical Typographer," Benjamin Livermore.

▶1866
"Machine for Writing and Printing," Abner Peeler.
    "Pterotype"; first practical typewriter of the typewheel class, John Pratt.

▶1867
First experimental, one-letter model, Christopher L. Sholes.
    First working model "Type-writer," Christopher L. Sholes.
    "Machine for Writing with Type or Printing on Paper or Other Substance," Thomas Hall.

▶1868
Ottmar is apprenticed to his uncle Louis Hahl in Bietigheim, who runs a watchmaker's shop.
    Movable type plate and manner; "Type-writer" (capitals only), Christopher L. Sholes.

▶1872
Ottmar emigrates to the United States. Ottmar joins the workshop of his cousin, August Hahl, in Washington, where electrical appliances and models for new inventions are built.

▶1874
August Hahl makes Ottmar his business manager. He meets several inventors and succeeds in acquiring the first patent of his own on March 17.
    First commercial typewriter (wrote capital letters only), Remington for Sholes and Glidden.

▶1875
August Hahl moves his workshop to Baltimore, Ottmar's contract with the inventors is not affected.

▶1876
Charles T. Moore brings the plans and blueprints of a kind of typewriter which is to produce a transfer of a page for printing by lithography.
    Remington No. 1 model—capitals only, Remington.

▶1877
The Moore machine is completed. The ideas for another machine is provided to Mergenthaler by the court reporter J.O. Clephane. He wants to have a matrix punching machine connected with the stereotype process.

▶1878
This apparatus sees completion at the end of this year. At the same time Ottmar goes into partnership with August Hahl.

Model 2 Remington (wrote capital and small letters for first time—by shift key and platen shift), Remington.

▶1879
Mergenthaler tries to improve the transfer machine but realizes that his efforts in this direction are doomed to failure, and stops all his experiments.

▶1881
Ottmar Mergenthaler marries Emma Lachenmayer. The marriage produces five children, a daughter and four sons.

▶1882
Mergenthaler separates from August Hahl and founds a factory of his own in Baltimore's Bank Lane.

▶1883
The attorney Hine from Washington supports Ottmar Mergenthaler with a substantial loan for the completion of his invention.

Rotary Matrix Machine. In this machine, finger keys controlled a rotary type wheel equipped with projecting characters. The characters were selected successively by the operation of the keyboard and indented a papier-mâché strip. The matrix strip thus formed was then cut up into lengths and secured to a flat backing sheet to form a page or column matrix. Justification of the lines was effected by crimping or cutting the matrix strips. Type metal was then cast into the assembled matrix strips and the printing plate obtained.

"Caligraph" (wrote capitals and small letters for first time with its first full keyboard ever made), Yost-Wagner.

▶1884
The spring of this year sees the completion of the first composing machine, which is exhibited to some friends of Mergenthaler's on July 26. This marks the first time that a slug is composed on a Mergenthaler machine.

First Band Machine. This machine, without metal pot, was equipped with a series of vertical bars, each carrying a full alphabet of type and space characters raised on its surface. The bars descended at the touch of a finger key, each bar being arrested to bring its selected character to a certain level. After the Line-o-type was assembled and justified, a papier-mâché strip was forced against it, thus producing a matrix for one line. The matrix strips were then assembled side by side to form a stereotype matrix and type metal was cast into it to form a printing plate.

Charles T. Moore and J.O. Clephane give up their plans for a matrix making machine.

▶1885

The second, improved composing machine is ready for operation in February and is exhibited in Washington in the Chamberlain Hotel.

Second Band Machine. This machine, with metal pot was the first to produce "lines-o-type" automatically, through the action of finger keys. Vertical bars containing an alphabet of female characters descended at the touch of a finger key, were brought to a common alignment and metal forced through a mold into the depressed characters in the bars, thus forming raised type on the front edge of the slug in the mold. The slug was ejected through trimming knives into a galley and the vertical bars were lifted to their original position, ready for the next line. A banquet is given honoring Ottmar Mergenthaler. The President of the United States, Arthur, and several high ranking officials praise the invention. But Mergenthaler still is not satisfied. He investigates the possibility of circulating matrices and develops blueprints for the "Blower" Linotype.

▶1886

The first "Blower" machine sets type for the *New York Tribune* on July 3rd. This was the first commercial line casting machine using small circulating matrices, each containing one character. The matrices, stored in vertical tubes, were released in the proper order by finger keys, delivered to an inclined chute along which they were carried laterally and successively by an air blast to form a composed line. This line was transferred to the face of a slotted mold, justified by wedge spacers and a slug was produced by forcing metal through the mold into the depressed characters of the matrices. The matrices were then lifted to the top of the machine and returned through a distributor to the vertical tubes.

The first commercial operation of this machine was in the office of the *New York Tribune* in July, 1886. The "Blower" also sets the first book ever printed without handset type: "The Tribune Book of Open Air Sports." Newspaper publisher Whitelaw Reid gives the machine the name "Linotype."

As a successor to the National Typographic Company, the Mergenthaler Printing Company is founded. The Linotype machines are built in series. Ottmar Mergenthaler installs a matrix factory and invents 30 special machines for the production of matrices.

▶1888

Up to February, 50 machines have been sold. Ottmar Mergenthaler disagrees with the company and starts a factory for the production of Linotypes all of his own. The blueprints for a new Linotype, subsidized by J.O. Clephane, are ready toward the end of the year and the first Simplex Linotype comes off the production line.

Ottmar Mergenthaler is struck with pleurisy.

▶1889

Mergenthaler constructs his last and best machine, the Simplex in its final form.

▶1890

The Simplex machine is exhibited in the Judge Building in New York City.

Square Base Model 1. This machine was the forerunner of the Linotype. Matrices were stored in an inclined magazine at top front of the machine. The magazine narrower than present-day magazines and accommodated matrices up to and including 11 point. It was the first machine to have a 90-character keyboard and to use matrices similar to the ones used today. The single magazine was removable from the rear of the machine.

The gas governor, the pot pump safety stop, the power-driven keyboard and numerous other improvements were incorporated in this model. Some of these machines were in use for fifty years.

The Mergenthaler Linotype Company is founded in Brooklyn. A Linotype factory is also founded in Manchester, England.

First portable typewriter (Ideal keyboard), Geo. C. Blickensderfer.

▶1892

1,000 Linotypes have been sold or rented by the end of this year.

Model 1. The massive square base of the original Model 1 was replaced by the star or regular base. The justification and other levers were made lighter and springs were used to perform the same function as the former weighted levers. This principle permitted the use of "automatics" so arranged that the machine would stop or the spring would expend without breaking any parts. The Model 1 was a commercially popular machine.

Ottmar Mergenthaler visits Germany and is reunited with his father.

▶1893

The Linotype is the sensation of the Chicago World Fair.

▶1894

On October 5, the first Linotype machine appears on the European continent. It is a British model and is scheduled to set "De Neederlandsche Financier" in Amsterdam. A protest strike of the newspaper employees postpones the first edition for five days.

Mergenthaler's associate, C.A. Albrecht, and the newspaper publisher William Mayer with his son Jaques arrive in Germany in November. At the same time the first Linotype ever to appear in Germany is shown in a small Berlin shop.

Ottmar Mergenthaler is struck with tuberculosis and begins a long period of treatment.

▶1896

The Mergenthaler Setzmaschinenfabrik GmbH. is founded in Berlin.

Ford typewriter. First practical machine of radial-striking-bar class.

First automatic ribbon reverse, Remington.

First "noiseless" typewriter (U.S.A. patent No. 567,241), W.P. Kidder-C.C. Colby.

▶1897
Ottmar Mergenthaler writes his memoirs in Deming, New Mexico. A prairie fire destroys all his belongings including his books and manuscripts. He returns to Baltimore and again goes to work on his writings.

▶1899
On October 28, Ottmar Mergenthaler dies in Baltimore. He is buried in the Loudon Park Cemetary.

▶1903
Model 2. This was the first two-magazine machine. Matrices from both magazines could be mixed at will in the same line and distributed automatically. It was the original mixer Linotype.

Selection of matrices from both magazines was controlled by a small lever which locked one escapement while unlocking the other. The lower magazine was shorter than the upper, was stationary and could only be removed for repairs. The upper magazine was wider than that of the Model 1, and was interchangeable with the magazine of Model 3. It was removable from the rear.

Model 3. A single magazine machine with a magazine two inches wider at the lower end than the one used on Model 1. In shape and size, the magazine was similar to the ones in use today. This Linotype was known as the "Pica" machine since faces pica (12 points) and over could be used. The faces previously used on the Model 1 had been limited to 11 point. The single magazine was removed from the rear of the machine.

▶1906
Model 4. A quick-change double magazine Linotype—a great advance over the Model 2 in that both upper and lower magazines were readily removable and changeable, the upper from the front—the lower from the rear. This was a mixing machine. Matrices from both magazines could be mixed in the same line and distributed to their respective magazines. A small lever at the right of the keyboard controlled the selection of matrices from each magazine, the magazines remaining stationary as in the

Model 2. The upper magazine was interchangeable with that of the Model 5.

Model 5. Quick change single-magazine Linotype. The light weight magazine was easily changed and was removable from the front of the machine. This was the first machine that could be used against a wall or backed up against another machine to save space. The escapement in former machines formed part of the magazine. In this machine the escapement was separated from the magazine when the magazine was removed. The magazines were wide enough to accommodate many 18 point faces and were interchangeable with the upper magazine of the double-magazine Model 4.

▶1911

Model 8. This was the first three-magazine Linotype. In its general features the Model 8 resembled the Model 5 but was a great advance over that model in that it carried three magazines, any one of which could be quickly brought into operation. The two upper magazines of this original Model 8 were readily removable from the front of the machine. The lower magazine could be removed but required a little more time. Along with the development of the Model 8, the Automatic Font Distinguisher was introduced—a device to prevent matrices from entering the wrong magazine.

Model 9. The first four-magazine Linotype. Radically new in many of its features, it was designed to meet varied requirements for advertising, display and job composition, calling for frequent change of face and body. It was a mixer Linotype and successfully solved the problem of mixing matrices from four magazines in one line and distributing the matrices automatically to their respective magazines. All magazines were removable from the front.

▶1914

Model 14. This was the first machine to have an auxiliary magazine. The addition of this 28-channel auxiliary magazine, located to the right of the main magazines, increased the range of the Linotype in the number and size of characters. The auxiliary magazine of this machine was operated by a separate punch keyboard to the right of the regular keyboard.

▶1918

Model 20. This was the first display Linotype. Originally made with one short magazine, it was later built with three short display magazines. The magazines contained 72 channels instead of the usual 90, the space between channels being wider to accommodate large faces. Display matter up to 30 point, and larger sizes of condensed faces were set like straight matter direct from the keyboard. The short magazines could be changed quickly and easily. This machine accomplished for display composition what the original Linotype did for news composition.

▶1920

First Standard keyboard portable, Remington.

▶1921

Models 21 and 22. These machines were an improvement over the Model 20. The Model 21 was a three-magazine machine. The Model 22 was the same as the Model 21 with the addition of a single auxiliary magazine. In addition to using the three magazines as 72-character magazines, they could be used as 55-character magazines by the simple swinging of a second channel entrance in place and a movement of a lever at the right of the keyboard. This permitted wider faces to run in 55 channels of the 72-channel magazine than was possible in the regular 72-channel layout, a valuable feature for newspaper heads.

▶1924

In Mergenthaler's birthplace, Hachtel, a memorial plate is affixed to the old schoolhouse, honoring the inventor on the 25th anniversary of his death. A small museum also is opened.

Models 25 and 26. These were mixing machines and were a distinct improvement over the Models 16 and 17 mixers brought out in 1916. The Model 25 carried two main magazines, and the 26, two main and two auxiliaries, brought into operating position by a convenient hand lever.

The Models 16 and 17 brought out in 1916, as mentioned above, were an improvement over the Model 9 introduced in 1911.

▶1935

Models 27 and 28. These were called Super-Display Linotypes

because their wide magazines accommodated normal 36 point and condensed faces up to 60 point. Model 27 was equipped with three wide 72-channel main magazines. Model 28 was similar to the 27, but supplemented the three main magazines with either one or three wide auxiliaries. The wide magazines were made of Linolite and were easy to change and shift. A swinging bracket held them conveniently for quick removal. For those few faces larger than the wide magazines or auxiliary magazines could carry, a detachable hand stick was furnished.

▶1937
First short stroke Standard typewriter, Remington Rand.

▶1938
Model 27, Remington Rand.

▶1945
The Mergenthaler Setzmaschinenfabrik GmbH. in Berlin is completely demolished by street fighting and bombing. The Russians destroy all remaining machines, matrices, and documents.

▶1949
Production finally gets underway despite huge difficulties.

▶1954
Honoring the 100th birthday of Ottmar Mergenthaler, a new museum is opened in Hachtel.

The three Linotype centers release the following overall production figures: Mergenthaler Linotype Company, Brooklyn, NY, 70,000 machines since 1890; Linotype and Machinery Limited, London, 15,000 machines since 1890; Mergenthaler Setzmaschinenfabrik GmbH. (Linotype GmbH.), Berlin, 13,000 machines since 1899.

BIBLIOGRAPHY

Annenberg, Maurice. *Type Foundries of America and their Catalogs.* Baltimore: Maran Printing Services, 1975.

American Dictionary of Printing and Bookmaking. New York: H. Lockwood & Co., 1894. Reprint. Detroit, MI: Gale Research Co., 1967.

American Printer, 1900-1925.

Bigmore, F.C., and Wyman, C.W.H. *A Bibliography of Printing.* Reprint. London: The Holland Press, 1969.

British Patent Office, London, copies of patents for typesetting machines, 1822-1925.

Burry, W. Turner, and Poole, H. Edmund. *Annals of Printing.* London: Blandford Press, 1966.

Butterworth, Benjamin. *The Growth of Industrial Art.* Washington, D.C.: Government Printing Office, 1892.

Diderot, Denis. *A Diderot Encyclopedia of Trades and Industry. Vol. 2.* Edited by Charles C. Gillispie. New York: Dover Publications, 1959.

Dreier, Thomas. *The Power of Print—And Men. Commemorating Fifty Years of the Linotype's Contribution to Printing and Publishing.* Brooklyn, NY: Mergenthaler Linotype Company, 1936.

Eckman, James. *The Heritage of the Printer.* Philadelphia: North American Pub. Co., 1965.

Hansard, Thomas C. *Typographia: An Historical Sketch.* London: Baldwin, Cradock, and Joy, 1825.

Hohne, Otto. *Geschichte der Setzmaschinen.* Leipzig: Der Deutschen Buckdrucker G.M.B.H., 1925.

Huss, Richard E. "*A Chronological List of Typesetting Machines and Ancillary Equipment—1822-1925.*" The Journal of Typographic Research 1 (1967): 245-274.

—*The Development of Printers' Mechanical Typesetting Methods, 1822-1925.* Charlottesville, VA: University Press of Virginia, 1973.

—*The Printer's Composition Matrix.* New Castle, Delaware: Oak Knoll Books, 1985.

Inland Printer 6-126 (1888-1950).

Kelber, Harry and Schlesinger, Carl. *Union Printers and Controlled Automation.* New York, NY: Free Press, 1967

Kubler, George A. *Historical Treatises, Abstracts and Papers on Stereotyping.* New York: George A. Kubler, 1936.

Lawson, Alexander S. *A Printer's Almanac.* Philadelphia: North American Pub. Co., 1966.

Legros, Lucien A. and Grant, John C. *Typographical Printing-Surfaces.* London: Longmans, Green and Co., 1916.

Lehmann-Haupt, Helmut. *The Book in America.* New York: R.R. Bowker, 1954.

Levine, I.E. *Miracle Man of Printing.* New York: Julian Messner, 1963.

Loy, William E. "*Designers and Engravers of Type.*" Chicago: The Inland Printer, Vols. 20-25, 1897-1900.

MacKellar, Smiths & Jordan. *One Hundred Years, 1796-1896.* Philadelphia: MacKellar, Smiths & Jordan, 1896.

Mackey, Joseph T. *The Linotype Machine and The Linotype Organization (Manuscript).* Unpublished: 1936.

McCorison, Marcus A. "*Dr. William Church of Vershire, Vermont.*" Printing and Graphic Arts, 4, no. 2: 35, 1956.

McMurtrie, Douglas C. *The Book.* New York: Covici Friede, 1937.

Mengel, Willi. *Ottmar Mergenthaler and the Printing Revolution.* New York: Mergenthaler Linotype Co., 1954.

Middleton, R. Hunter. *Making Printers' Types.* Chicago: The Black Cat Press, 1938.

Milestones of Machine Typesetting. New York: Mergenthaler Linotype Company, 1944.

New Wings for Intelligence. Baltimore: Schneidereith & Sons, 1954.

Reed, Talbot B. *A History of the Old English Letter Foundries.* Edited by A.F. Johnson. London: Faber and Faber Ltd., 1952.

Ringwalt, John L. *American Encyclopedia of Printing.* Philadelphia: Menamin & Ringwalt, 1871.

Roby, Henry W. *Henry W. Roby's Story of the Invention of the Typewriter.* Menasha, WI: George Banta Publishing Co., 1925.

Rusting, Robert R. "*The Watchmaker's Apprentice Who Invented the Linotype.*" Watch Repair Digest, April, 1957.

Schoenrich, Carl Otto. *Biography of Ottmar Mergenthaler and History of the Linotype, Its Invention and Development.* Baltimore: The Friedenwald Company, 1898.

Schori, Ward K. *Print Shop Nostalgia.* Evanston, IL: The Schori Press, 1962.

Scientific American, 1845-1920.

Scully, Michael. "*Mergenthaler's Wonderful Machine.*" The Reader's Digest, March, 1953.

Senefelder, Alois. *The Invention of Lithography* (Translated from the original German by J. W. Muller). New York: The Fuchs & Lang Manufacturing Company, 1911.

Sherman, Frank M. *The Genesis of Machine Typesetting.* Chicago: M & L Typesetting and Electrotyping Co., 1950.

Silver, Rollo G. *Typefounding in America, 1787-1825.* Charlottesville, VA: University Press of Virginia, 1965.

The Story of the United States Patent Office, 1790-1956. Washington: U.S. Department of Commerce, 1956.

Thompson, John S. "*Composing Machines—Past and Present.*" Inland Printer 30-32 (Oct. 1902-Nov. 1903).

—*History of Composing Machines.* Chicago: Inland Printer Co., 1904.

Tracy, George A. *History of the Typographical Union.* Indianapolis, Indiana: International Typographical Union, 1913.

United States Patent Office, Washington, D.C., copies of patents for typesetting machines, 1854-1913.

Updike, Daniel B. *Printing Types, Their History, Forms and Use.* Cambridge, MA: Harvard University Press, 1937.

Usher, Abbott Payson. *A History of Mechanical Inventions.* Cambridge, MA: Harvard University Press, 1954.

Woodcroft, Bennett. *Abridgements of Specifications Relating to Printing.* London: 1859. Reprint. London: Printing Historical Society, 1969.

Wroth, Lawrence C. *The Colonial Printer.* Charlottesville, VA: University Press of Virginia, 1964.

INDEX

Amalgamated Typesetting Machine Company, 102
American Type Founders (ATF), 82, 99
Arthur, Chester A., 50
Bancroft, J. Sellers, 84
Benton, Linn Boyd, 68, 81-82
Binny & Ronaldson, 99
"Blower Linotype", 85
Bullock, William, 33
Clephane, James Ogilvie, 7-8, 23-28, 30-32, 35-37, 40-43, 46-47, 50, 57, 65, 79, 85-86, 92, 101, 105
Clough, Jefferson, M., 11
"Church machine", 34
Crandall, Lucien Stephen, 14
Crossman, J.H., 37
Densmore, James, 6-7, 10
Dodge, Phillip Tell, 14, 81, 89-96, 100, 105
Edison, Thomas Alva, 10, 96, 104
DeVinne, Theodore L., 97-98
Ged, William, 29
Glidden, Carlos, 3, 14
Greeley, Horace, 58, 63
Gutenberg, Johann, 33, 45
Hahl, Louis, 18-20
Hahl, August, 19-20, 25, 32, 36-40, 46, 49, 78
Haldeman, W.N., 58
Hammond, James B., 15
Hine, L.G., 39, 43, 47, 50, 55-56, 79, 86
Hoe, Richard, 33
Hutchins, Stilson, 49, 56, 58

"Impression machine", 30-32
International Typesetting Machine Company, 103
Intertype Corporation, 103
Jenne, William K., 11
Lanston, Tolbert, 84-85
Lawson, Victor, 58
"Linotype", 64, 66, 100, 104
Locke, D.R., 11, 13
Mackey, Joseph T., 89
Mergenthaler Linotype Company, 87, 89, 100
Mergenthaler, Ottmar, 17-19, 20, 28-28, 30-32, 36-38, 39-43, 46-47, 57-58, 63, 66-68, 72-73, 79, 86-87, 103-104
Mergenthaler Printing Corporation, 58, 63, 78-79, 85, 88-89, 91
"Monoline", 102
Moore, Charles T., 8, 21-25, 26-28, 32
Muehleisen, Carl, 101
Nasby, Petroleum, 11, 13 (Also see Locke, D.R.)
National Typographic Company, 44, 55, 58, 73, 76-77, 88, 91
New York Tribune, 63, 66-67, 104
"Paige Compositor", 82
Papier-mâché, 29-32
"pterotype," 3
Rand, W.H., 58
Reid, Whitelaw, 58, 60-64, 65-67, 69-78, 85, 88, 105
Remington, 11-14, 90

Ridder, Hermann, 102-103
Rogers, J.R., 87, 90
"Rotary Impression Machine", 32
Schoenrich, Carl Otto, 101-102
Seamans, Clarence Walker, 14
Sholes, Christopher Latham, 1-16, 26
Schuckers, J.W., 87
Scudder, W.S., 94, 103
"Simplex Linotype", 78-79
Smith, Henry, 58
Soulé, Samuel W., 3, 14
Stereotyping, 29, 32
Stone, Melville, 58-62
Twain, Mark, 11-13, 82-84
Typefounding, 45, 54, 99
Typewriter, 4-6
"Typograph", 87-88, 90
Underwood, John T., 15
Varityper, 15
Waldo, Robert V., 81
Westinghouse, George, 34
Wyckoff, William Ozmun, 13
Yost, George Washington, 10-11, 14

COLOPHON

Sholes and Mergenthaler (and mostly Clephane) would have been pleased with the production of this book. The original manuscript was typed on a personal computer using Xywrite III word processing software—an extension of the typewriter—which permits electronic editing. A proof of the pages was printed out on a laser printer, which in some cases could have been used for final output. That combination of technologies appears to be bridging the gap between the typewriter and the printed page.

The floppy disks were sent to Robey Graphics in Washington, DC and converted to the coding of their computer typesetting system, which was programmed for automatic kerning, ligatures and hyphenation/justification. Copy was then output on a Linotron digitized phototypesetter made by The Linotype Company. Digitized type is set as patterns of dots, which is as far away from hot metal slugs as you may ever get.

The text was set in Goudy Old Style, a typeface designed by Frederic Goudy, who did most of his work for the Monotype.